T0331257

"Infrastructure is perhaps the fastest growing asset class, but it remains relatively immature and poorly understood. This book provides valuable insights into evolution of the asset class, the perspectives of the main stakeholders and the opportunities and challenges for the future. Over the past 15 years, Infrastructure has emerged as an important asset class for institutional investors. Despite this, the asset class remains relatively immature with many aspects poorly understood. In this context this book provides valuable insights into infrastructure investing, from the perspectives of investors, governments and, most critical of all, the users and beneficiaries of the asset class."

Dr Peter Hobbs, *Managing Director, Head of Private Markets, bfinance*

"James' book is a valuable contribution to the growing literature on infrastructure businesses. Successfully utilizing private capital is an imperative for governments looking to build and maintain their evolving infrastructure needs and this book enriches our understanding of what the private sector seeks out when assessing infrastructure investments. Public stakeholders, students, industry professionals and anyone else keen for infrastructure will enjoy reading this book."

Aaron Vale, CFA, CAIA, *Managing Director, Co-Head of Indirect Infrastructure, CBRE Investment Management*

"Given the growing realisation that the quality of infrastructure is absolutely key to every country's aspirations, James McKellar's work is not only timely but also spot on: just as much as there is an 'infrastructure gap', there is a 'knowledge gap' between the public and private on how to successfully prepare, procure, fund, finance, operate and maintain the type of future-ready infrastructure people expect. McKellar's work makes great strides in helping to close that gap."

Matthew Jordan-Tank, *Director, Sustainable Infrastructure Policy and Project Preparation, European Bank for Reconstruction and Development (EBRD)*

"James' perspective on the flow of capital into infrastructure is both refreshing and needed. Understanding how to access private sector creativity and risk-taking is crucial to addressing the most vexing problems in infrastructure today. The challenge to the public sector is to find ways to harness the natural profit-seeking motivation of the private sector in the service of the public interest. The potential exists cross the globe and the returns to both the public and private sector can be immense and simultaneous."

Stephen C. Beatty, *Global Chairman (Non-Exec), Infrastructure and Chairman, Global Cities Center of Excellence, KPMG in Canada*

Infrastructure as Business

Infrastructure as Business brings new emphasis and clarity to the importance of private investment capital in large-scale infrastructure projects, introducing investors, policymakers, and other stakeholders to a key element that is surprisingly absent from the discourse on public–private partnerships. Despite the importance of modernizing infrastructure across the globe, governments often face challenges in securing the necessary capital to meet future need, as well as developing policy to meet these goals. Explaining the structure of the private investment universe and flow of private capital in such projects, this book ambitiously aims to bridge this "infrastructure gap" by elucidating shared terminology, conceptual frameworks, and an alignment of goals and objectives between public and private sectors—essential to meet increasing environmental, social, and governmental requirements for infrastructure in coming years.

- Appropriate for graduate-level courses in real estate, public policy, and urban planning that focus on infrastructure, project finance, and procurement and delivery models such as PPPs.
- Provides a clear understanding of private investment and PPPs to the investment community as well as professionals in real estate, project finance, and related fields, who often learn mostly on-the-job and from colleagues.
- Equips government officials and policymakers with key terms and concepts needed to "sit across the table" with private financers and explore opportunities for private capital investment in early project stages.
- Outlines communication strategies for both public and private sectors, which will increasingly need to collaborate to address climate change, respond to new technologies, and develop efficient ways to deliver services.

Written to engage academic, private investment, and public policy/governance audiences alike, *Infrastructure as Business: The Role of Private Investment Capital* invites discussion and opens doors to advancing new business models, with international applications, to offer increased value for private investors as well as more efficient, flexible funding for innovative infrastructure development in the future.

Infrastructure as Business
The Role of Private Investment Capital

James McKellar

Routledge
Taylor & Francis Group

NEW YORK AND LONDON

Designed cover image: Clelia Iori

First published 2024
by Routledge
605 Third Avenue, New York, NY 10158

and by Routledge
4 Park Square, Milton Park, Abingdon, Oxon, OX14 4RN

Routledge is an imprint of the Taylor & Francis Group, an informa business

Library of Congress Cataloging-in-Publication Data
Names: McKellar, James, author.
Title: Infrastructure as business: the role of private investment capital / James McKellar.
Description: New York, NY: Routledge, 2024 . |
Includes bibliographical references and index.
Identifiers: LCCN 2023009311 (print) | LCCN 2023009312 (ebook)
Subjects: LCSH: Infrastructure (Economics) | Public works—Finance.
Classification: LCC HC79. C3 M45 2024 (print) |
LCC HC79.C3 (ebook) | DDC 363—dc23/eng/20230320
LC record available at https://lccn.loc.gov/2023009311
LC ebook record available at https://lccn.loc.gov/2023009312

ISBN: 9781032501161 (hbk)
ISBN: 9781032493176 (pbk)
ISBN: 9781003396949 (ebk)

DOI: 10.1201/9781003396949

Typeset in Times New Roman
by Newgen Publishing UK

Dedicated to

James McKellar

1906–1964

Contents

Figures

Acknowledgments

This book represents a synthesis of knowledge acquired in the last decade of my academic career, a career that spans architecture, real estate, and now infrastructure. All three reflect a life-long interest in shaping the built environment and an appreciation of the effort it takes to accommodate human settlement. These settlements represent the nexus between real estate and infrastructure and are about *people* and *things*. Infrastructure attracted my attention in the early 2000s at a time when public–private partnerships (PPPs) were promoted as a procurement model for the delivery of infrastructure. However, PPPs offered a very narrow perspective that focused on the procurement and financing of new public infrastructure. Thinking took place in silos, and people were never mentioned as a key part of the process by which we build, finance, and operate infrastructure. Nor did the users get much mention. The focus failed to acknowledge the importance of private investment in, and ownership of, infrastructure across the globe.

It was a comment by Ron Mock, then CEO of Ontario Teachers, that launched me in a particular direction when thinking about infrastructure. From the perspective of this successful institutional investor, infrastructure was a business. He explained why some of the largest private investors in infrastructure were so successful, particularly a number of Canadian pension funds. Shortly thereafter, he approached me to ask if the Schulich School of Business at York University would partner with a group of industry leaders to launch a Fellowship program. The Investors Leadership Network (ILN) was comprised of 13 of the world's largest institutional investors in infrastructure and their goal was to assist emerging economies in launching sustainable infrastructure projects that could be, in his words, "bankable". What I learned during the first three years of the Sustainable Infrastructure Fellowship Program served as the backbone for this book.

All of this would never have happened without the continued support from Dean Dezso Horvath of the Schulich School of Business, now Dean Emeritus, who was unwavering in his support to integrate the business side of infrastructure into our graduate real estate program. This shift also emphasized the importance to investors of real assets that encompass both real estate and infrastructure.

The focus on private investments in infrastructure can be traced back to the three founders of the ILN. These gentlemen were responsible for the Fellowship and opening a window on an industry sector that academics have limited access to. Marc-André Blanchard was then Canada's Permanent Ambassador to the United Nations (2016–2020) and now Executive Vice-President and Head of CDPQ Global and Global Head of Sustainability; Michael Sabia was then President and CEO of CDPQ (2009–2020) and is now Deputy Minister of Finance in the Government of Canada; and Ron Mock who retired as President and CEO of Ontario Teachers in December 2019. All three brought to my research a broad definition of infrastructure, an understanding of why investing in infrastructure is a global business, why emerging markets matter, and the importance of sustainable infrastructure. They also shed light on why a group of Canadian pension funds dominated the institutional investment market in infrastructure, across the globe.

I started with a round of interviews with individuals on the front lines of some of the leading pension funds. A number of common themes emerged that reinforced my commitment to the book. I was then exposed to the unlisted infrastructure funds at a conference in Berlin in 2019, a conference held each year that could be characterized as an annual platform for private deal-making. The conference reinforced the importance of private investment funds flowing into infrastructure from a large cadre of relatively small international firms. I also tracked the meteoric rise of Toronto-based Brookfield, the second largest listed fund in the world, after Blackstone. It was also consequential that the largest pension fund investor in infrastructure, CPP Investments, was Canadian based, as were Caisse de dépôt et placement du Québec (CDPQ), Ontario Teachers' Pension Plan (OTPP), Ontario Municipal Employees Retirement System (OMERS), PSP Investments, and AIMCo. What accounted for this Canadian dominance and their acknowledged success? A question that remained forefront in my research.

A major source of information was the Industry Day presentations in the Fellowship program in which international investors gave a short presentation on their operations and presented an investment case. Two industry representatives were responsible for the Industry Days, Stacey Purcell, Managing Director, Infrastructure and Natural Resources, and Head of Latim (Latin American Team for Infrastructure) at OTPP, and Eve Bernèche, Managing Director, Infrastructure at CDPQ. I am indebted to both women for much of what I learned about the industry. Two of my colleagues were with me on this journey of learning; Karen Shlesinger, Program Director of the Fellowship and Sherena Hussain, Academic Director of the Fellowship. I thank both of them for what they contributed.

I must also thank the Fellows themselves who never stopped asking intelligent questions, were eager to talk directly with investors, and offered me a window on a part of the industry that few would have the opportunity to explore. The Fellowship reinforced the need for this book.

Finally, I must acknowledge the contribution and support from my wife, Clelia Iori, who also designed the cover for the book. She reviewed the early drafts and lent a critical eye as I assembled the chapters. Nor did she let me forget the importance of sustainability in this exercise.

Author biography

Professor Emeritus James McKellar, Brookfield Centre in Real Estate and Infrastructure, Schulich School of Business, York University, was Associate Dean, External Relations, and former Director of the Brookfield Centre. Prior to joining York University, he was a faculty member at the Massachusetts Institute of Technology (MIT) and served as the first Director of the Center for Real Estate at MIT. In addition to MIT, he held faculty appointments at the University of Pennsylvania and the University of Calgary.

Professor McKellar has a life-long involvement in the real estate industry and has consulted to businesses and governments in many parts of the world on real estate matters covering housing, development, finance and investment, asset management, and market performance. His recent work focuses on infrastructure in the context of private investment, rapid urbanization, population growth, and sustainable infrastructure, particularly in emerging economies where new technologies, new business models and socio-cultural factors play a significant role. He is actively involved in the Sustainable Infrastructure Fellowship Program announced in June 2018 to coincide with the G7 held at Quebec City, Canada. He has been involved in strategic issues related to real estate and infrastructure in both the private and public sectors. As principal of McKellar Associates Ltd., he specialized in strategic issues related to real estate and infrastructure.

Introduction

Few books about infrastructure have been produced in the last decade (Gómez-Ibáñez and Liu, 2022). This is surprising given the importance of infrastructure in the world today. Infrastructure is a significant part of the foundation upon which society supports its social and economic needs and is essential to providing for the basic needs of shelter, clean water, sanitation, food, and fuel. It is the physical manifestation of contracts between a government and a society over which it rules, between a government and its industrial base, and between countries in a highly competitive global environment. Infrastructure influences how we move through and between cities, communicate with each other, deliver health care and education, connect sources of production to points of consumption, pursue innovations, and move people out of poverty. Infrastructure is intimately connected to urbanization, global connectivity, and the prosperity of nations. In the last decade infrastructure has been elevated to a prominent role in addressing climate change and securing the sustainability of the planet. The paucity of literature on infrastructure is even more surprising given the extremely poor record of most governments in meeting their own objectives for improved and new infrastructure. Even in advanced economies, infrastructure today gains more recognition for its absence, or poor state of maintenance and repair, than it does for advancing the goals of society.

This book adds a dimension to the complex topic of infrastructure that is absent from most of the current literature dealing with infrastructure and may explain some of the challenges and uncertainties that exist. This dimension addresses the expanding role of private investment capital in infrastructure globally. In addressing private investment in infrastructure, one must recognize a definition of infrastructure that extends well beyond what is commonly recognized in the public sector as infrastructure. The book examines the basis upon which private investors raise and spend billions of dollars annually to deliver infrastructure services, many of which governments recognize as essential to the social and economic well-being of their respective nations. The book explores a side of infrastructure not well understood, particularly by governments, namely the ownership of and private investments in infrastructure that represent businesses spanning a broad spectrum of the economy.

DOI: 10.1201/9781003396949-1

Private ownership and operation of infrastructure is less transparent and excluded from many studies on this topic. In comparison to public infrastructure, private investment in infrastructure lacks comprehensive data sources, may not fit a narrow definition of infrastructure applied in the public sector, and is often seen as having no clear role in serving a public purpose. A railroad track may get people to and from their desired destination and, despite being privately owned and operated, the public need is clear. Same for an airport. When used for hauling cargo the public interest is less clear, although governments recognize the economic benefits of moving goods and materials. The public sector seldom sees infrastructure as a revenue generator, other than to cover the cost of the service being delivered. Infrastructure is often free of any user charges such as when accessing roads and highways other than a toll road, or entering a school, courthouse, hospital, or library. In comparison, private investors must secure an ongoing income stream to cover costs, plus generate a profit. It is profit that motivates those to take on business risks. What distinguishes investment activities in infrastructure within the private sector are the decisions made from a business perspective that include investing in people who run these businesses, as well as in the underlying assets.

Most books and published articles on the topic of infrastructure released over the past two decades offer a public sector perspective and focus on the design, delivery, and financing of public infrastructure, with an emphasis on social infrastructure and based on a narrow definition of what constitutes infrastructure. Topics are heavily weighted to public–private partnerships (PPPs), public policy, project finance, project management, and defining optimal public infrastructure spending. Early contributors on the topic of public–private partnerships include Darrin Grimsey and Mervyn K. Lewis (2007) on infrastructure provision and project finance; Mirjam Bult-Spiering and Geert Dewulf (2008) dealing with strategic issues from an international perspective; E.R. Yescombe and E. Farquharson (2018) on PPPs and principles of policy and finance; Akintola Akintoye, Matthias Beck, and Mohan Kumaraswamy, Editors (Akintoye et al., 2015) on policy and finance for PPPs; Hodge, Greve and Boardman, Editors (2010) have a handbook on PPPs; and Jeffrey Delmon (2011) provides a guide on PPPs for policy makers.

There are also a number of books that deal almost exclusively with project finance. Books on this subject include E.R. Yescome's early work on the principles of project finance (2002); John D. Finnerty (2013) addressing asset-based financial engineering; Stefano Gatti (2018) addressing the theory and practice of project finance; Neil S. Gregg (2010) details the infrastructure sector; and Rajeev J. Sawant (2010) addresses the management of risk and rewards of investing in infrastructure for pension funds, insurance companies, and endowments. This book is not about PPPs. In fact, the emphasis on PPPs is somewhat surprising given the estimate that PPPs represent somewhere between 3 and 5 percent of new infrastructure projects, the remainder being delivered through traditional procurement processes.

A series of papers published in a special edition of the *California Management Review* (Gil and Beck, 2009) provide an excellent overview of issues in the provision of infrastructure, including bridging infrastructure with business and management, PPPs, innovation and technology, strategic planning for the infrastructure sector, and an analysis of PFIs in the UK applied to hospital programs. A book that more closely aligns with the objective for this book is that of Barbara Weber and Hans Wilhelm Alfen (2010) that deals with infrastructure as an asset class and includes investment strategies, project finance, and PPPs. An important contribution that broadens the discussion is that of Jonathan Haskel and Stian Westlake (2018) who introduce the intangible economy and the relationship to infrastructure. Two recent books on infrastructure by noted academics are those of José A. Gómez-Ibàñez and Zhi Liu (2022) dealing with infrastructure economic and policy from an international perspective, and Edward L. Glaeser and James M. Poterba, Editors, (2022) on economic analysis and infrastructure investments.

Unfortunately, there is limited information on those in the private sector who invest billions of other people's money in infrastructure, other than what is contained in trade publications and presented at industry conferences. Private investors are not large sources of infrastructure financing, they are entrepreneurs relying primarily on sources of equity raised through various fund structures. Many infrastructure investment funds, being private, are protective of their confidentiality as they operate in a highly competitive market and compete on the basis of derived fund performance and fees charged. It is very difficult to get information on how these funds operate, how they make decisions, and the business culture within which they control their general partner (GP) and limited partner (LP) structures. What is known is heavily weighted to the use of data compiled across funds by several industry publications and indices, with a distinct focus on investment yields.

Investment funds may be involved in project finance for some of transactions, but to a large extent they are equity investors, relying upon forms of financial analysis and cash-flow modeling not dissimilar to that used in the corporate world. There are three distinctions that set the financial analysis of infrastructure apart from that of real estate, the other real asset likely found in an investment portfolio: comparatively long time period; an emphasis on both quantitative and qualitative factors given that it is the combination of a fixed asset and the operation of the asset that must be assessed; and the importance of crafting a portfolio of investments to achieve the desired risk/return relationship for a given fund. An understanding of corporate finance may be the more appropriate basis for making what are essentially business decisions in allocating private funds to infrastructure[1].

The idea for the book originated from two sources. The first was a conversation back in early 2018 with Ron Mock, then CEO of Ontario Teachers' Pension Plan (OTPP), one of Canada's largest and most successful pension plans. He reiterated comments he made as a speaker at the Milken Institute Global

Conference in Los Angeles in May of 2017 emphasizing that pension managers in the U.S., the UK, Asia, and elsewhere run their investment strategies "as acquiring businesses". By investing private equity in infrastructure, either on their own, or in consortia or partnership arrangements, investors are pursuing business opportunities. This investment strategy has been refined by large pension funds, global sovereign funds, and large insurance companies, OTPP among them. Ron Mock claimed that OTPP *"cares less about the nomenclature of alternatives"*—meaning how asset classes are defined—than the value proposition of a deal. Thinking of infrastructure as a business proposition seemed radical in comparison to the messages delivered in books and publications on infrastructure. The challenge was gaining sufficient insight into this burgeoning investment community to assemble enough material to support a book.

The launch of the Sustainable Infrastructure Fellowship Program in 2018 by a group of like-minded investors opened a window on institutional investors that few academics have access to. This announcement coincided with the G7 meeting in Quebec City, Canada, that year. Under the initiative of three individuals, Marc-André Blanchard, then Canada's Permanent Representative to the United Nations and now a senior executive with Caisse de dépôt et placement du Québec (CDPQ) pension fund in Quebec; Ron Mock, then CEO of OTPP and now retired, and Michael Sabia, then CEO of CDPQ and now Deputy Minister of Finance in the Government of Canada, a select group of large investors, dominated by pension funds, established the Investors Leadership Network (ILN), an organization that currently has 13 members, many of whom are among the largest investors in infrastructure globally.

The Schulich School of Business at York University in Toronto, Canada, was selected as the ILN's education partner for the Fellowship program. The intent of the Fellowship is to provide senior government officials in emerging economies involved in infrastructure with a better understand of what private investors are looking for, how they make investment decisions, and how government can make infrastructure projects more "bankable" to investors. The Fellowship opened a direct line of communication between senior government officials in emerging economies and a group of global investors in infrastructure willing to share information, experiences, and expectations. The insights gained through access to these large investment funds were invaluable in formulating this book. These insights allowed the book to move beyond theory, explore an industry sector not well understood despite its size and global reach, and provide information that could contribute to a more informed approach to harnessing private capital to invest in infrastructure and address sustainability mandates.

Comparing the involvement of infrastructure investors to that of governments, led to a series of questions that helped frame the approach to this book, questions that appear relatively simple but belie the complexity of this topic. Addressing these questions was like peeling the layers of an onion—each layer uncovered more questions than answers. Questions ranged from the most basic: What is infrastructure, to the more complex. Why is the topic of ownership of infrastructure such a "hot button" on each side? Then there are the more obvious questions. What is the infrastructure gap that is so widely cited? With so much private capital available and committed to infrastructure, why is there an infrastructure gap? Given the magnitude of need, why do investors see the real challenge is not raising money, but spending the money they already have. Some questions raise issues not always easy to comprehend. Why do "privatization" and private investment schemes raise concerns even among developed nations? Why do investors avoid public–private partnerships when governments promote them as an effective way to partner with the private sector? How do inventors decide what to invest in?

The book ventures into the details on how private capital flows into infrastructure to better understand the structure of this investment community and how they approach transactions. Through what conduits do investors channel their capital flows? What do investors mean when they say they are investing in a business when they acquire an infrastructure asset? What returns do they seek?

From a market perspective, there are clear patterns that define investment preferences, and these don't necessarily correlate with where governments might see their greatest needs. Examples are the avoidance by most investors in greenfield projects and particularly public–private partnerships, despite the urgency to launch new infrastructure projects by governments. This does not preclude investing in new infrastructure to support corporate and business needs, and where counterparty risks can be minimized.

Are there investment structures investors prefer? What infrastructure sectors do investors gravitate toward, and why? Can investors be encouraged to invest in new infrastructure projects and be permitted to transform the design, development, and operation of new infrastructure into an operating business? What is the potential for investing in the emerging economies where future demand will be strong? Behind such questions is an overarching interest by both public officials and private investors in pursuing what they both refer to as business models. What is meant by a business model? Are there new business models that might promote a more rewarding mutual relationship between public and private interests than public–private partnerships have managed to achieve to date.

Any discussion on new business models brings to the forefront the sustainability challenge. Private investors are embracing sustainability, despite the lack of clarity on what this really means. At one level there are increasing market pressures to meet ESG requirements in making investment decisions, but many ESG objectives are vague and controversial. Are the three letters in this acronym weighed equally? Investors are keenly aware of the impacts of climate change and commitments to achieve net-zero targets on their long-term holds, but could ESG requirements negatively affect the bottom line? Will pressures by governments to meet the UN Sustainability Development Goals (SDGs) shift investment preferences in the decade ahead?

Searching for answers to these questions, and more, contributed to the content of this book. The intent is to seek possible ways to bridge two divides, that of the public and the private realms, and see if this can be achieved from a perspective of infrastructure as a business. Within the public realm there is a call for increased private investment in infrastructure, but confusion in understanding what governments mean when they issue this call. There is also the limitation that infrastructure is seldom recognized by business and management studies beyond matters of public policy, the impacts of ESG on investment decisions, and climate-related finance. Infrastructure has gained from studies in planning, engineering, project management, finance, and project management. Surprisingly, business schools have largely avoided the topic of infrastructure despite the size and importance of the industry. They are beginning to embrace the topic of sustainable finance and explore the ramifications of ESG on investment decisions. Hopefully, their interests will expand as the pressures for new infrastructure increase and private investors assume a more prominent role in the search for new infrastructure solutions and ways to contribute to the sustainability initiatives.

It would be useful to clarify a few terms that are used throughout the book and are associated with buying and operating something to make money, namely *assets* and *investments*. These are confusing terms and used in a variety of ways (Haskel and Westlake, 2018). The same applies to the terms *funding* and *financing* that are typically used interchangeably and yet have very different meanings. According to the UN's System of National Accounts (SNA, 2008), "*investment is what happens when a producer either acquires a fixed asset or spends resources (money, effort, raw materials) to improve it*"[2]. SNA (2008) defines an *asset* as an economic resource that is expected to provide a benefit over a period of time. A *fixed asset* is an asset that results from using up resources in the process of its production, as distinguished from a *financial asset* such as buying shares in a publicly traded company. To be deemed an investment, the entity doing the investing has either to acquire the asset from somewhere else, or incur some costs to produce the asset themselves. This infers *spending resources.*

When addressing "*investment*" in infrastructure, this book is about spending by businesses to create fixed (non-financial) assets—*spending resources* to create a long-lived stream of productive services that can be called "capital" (Haskel and Westlake, 2018). It is this capital, combined with labor, that produces the infrastructure services we require. The people who make these fixed assets productive are an essential part of the infrastructure equation.

Organization of the book

The book is organized in six parts, plus an Introduction and Conclusion. Part I, Setting the Context, provides an overview of infrastructure, past, present, and future, and sets the framework in terms of outlining key issues that are addressed in subsequent chapters. An historical perspective is an essential part of any discussion on infrastructure given the role it has played from shaping early settlement patterns, to supporting military conquests, to nation and city building, improving the health and welfare of societies, and promoting new technologies and innovations leading to new infrastructure solutions. History can shed light on the origins of some of the ideological and political differences that prevail today, and the changing patterns of public versus private owner- ship of infrastructure over the centuries.

Part II, Investing in Infrastructure, outlines the growing appeal of real assets to investors. This contributes to understanding how private capital approaches infrastructure investments and the basis of the value proposition driving investment decisions. This builds on the case for infrastructure as a business proposition. Part III, The Investment Universe, covers the scope, size, and players in the infrastructure investment market and their performance expectations for enhanced yields, versus other investment options. This is a universe that has evolved rapidly and with investment structure that addresses demand and supply factors, and directs capital through various conduits and investment platforms. Part III offers perspectives on the investment industry as well as on public finance challenges.

Part IV, Sustainable Infrastructure, explains what sustainable infrastructure involves, discusses the impacts of climate change, the importance of building sustainable infrastructure, the rise of climate-related finance, and the emphasis on ESG that is influencing investment decisions and imposing new steward- ship requirements. Part V, Investing in New Infrastructure, discusses the invest- ment risks involved in greenfield projects and the challenges of project delivery, plus the role of PPPs in meeting future infrastructure needs. This section also addresses opportunities in emerging and frontier markets and summarizes some of the lessons learned with involvement in new infrastructure projects.

Part VI, New Frontiers for Investors, deals with the rise of intangible infrastructure assets, covers some of new investment opportunities that are shaping tomorrow's infrastructure, and the need for creativity and innovation in addressing future needs. Concluding remarks summarize the key messages in the book.

An increasingly uncertain world

When this book was conceived back in 2018, it was a different world with much more certainty in the investment world and greater confidence in global cooperation than exists today. Even discussions on the potential impacts of cli- mate change were beginning to gain traction. In the intervening period, major events have occurred that have global significance and will impact infrastruc- ture in ways that are not yet known. The COVID-19 pandemic, and its sub- sequent variants, are having long-term effects and there is recognition that a return to a pre-pandemic world is no longer a possibility. The pandemic has already forced the reconsideration of many of the traditional assumptions underlying infrastructure, whether dealing with energy, transit, airports, health and education, logistics, and the where and how people work.

The war in Ukraine is escalating and unfortunately it is the destruction of infrastructure throughout Ukraine that is grabbing the headlines and affecting the most vulnerable. It is not clear if the ambition of President Vladimir Putin is to re-create a Russian sphere of influence and establish boundaries for a new Russian empire. Coupled with this is China's quest for regional and potentially global primacy under President Xi Jinping. His threats to increase global competition and increase confrontation with the U.S. on trade and technology issues are cause for concern. These geopolitical events must be seen in the context of higher interest rates, inflation, and a recession. Nationalism is also on the rise. None of these portend well for addressing infrastructure needs, many of which stretch across borders. The same may be said for addressing climate change. With disorder on the rapid rise, the world agenda has been transformed in less than 12 months. None of this is dealt with in the book, nor can it be at this stage in a series of yet unfolding global events.

Hopefully, things will turn out for the better. If this prospect is realized, infrastructure will be more important than ever to ensuring a brighter future and bridging the collision between old thinking and new challenges. It will also be a test for principles of democracy and the role of private capital in contributing to a better world.

Notes

1 An example would be *Principles of Corporate Finance, 8th Edition.* (2016). Brealey, Richard, Stewart Myers and Franklin Allan. McGraw-Hill/Irwin. A recommended text on real estate finance and investment is that of Geltner, David M., Norman G. Miller, Jim Clayton, and Piet Eichholtz. (2016). *Commercial Real Estate Analysis and Investment*. Thompson Southwestern.
2 The System of National Accounts (often abbreviated as SNA; formerly the United Nations System of National Accounts or UNSNA) is an international standard system of national accounts, the first international standard being published in 1953. Handbooks have been released for the 1968 revision, the 1993 revision, and the 2008 revision. https://en.wikipedia.org/wiki/System_of_National_Accounts

Part I

Setting the context

1 Defining infrastructure

Infrastructure means different things to different people. Civilizations cannot function without infrastructure that provides clean water, a secure food supply, basic shelter, a reliable energy source, transportation, communications, waste management, education, and healthcare services. Infrastructure represents the requisite support systems for economic activity, the delivery of social services, expanding the economic base, enhancing productivity, and hopefully making cities more livable for all residents.

Infrastructure serves both public and private interests. A shared understanding is that infrastructure has a physical presence, but today that assumption no longer holds true with the rise of the digital economy and the virtual realm. Despite the widespread reference to infrastructure by politicians, professionals, industry leaders, and the media, it is not well understood what the term applies to. Not that infrastructure is something new. Infrastructure has a history dating back centuries, existed in the accoutrements and logistics of military campaigns, and was essential to the armadas and transportation networks in support of conquests through the centuries. Trade routes, ports, roads, and railway lines gave rise to villages, towns, and eventually cities, all of which require infrastructure. In modern times, infrastructure provides the foundation upon which businesses are launched and operate, whether serving local markets or all parts of the globe.

Modern-day infrastructure harkens back to an era of great construction enterprises and entrepreneurial ventures including the building of utilities, bridges, power generation and distribution facilities, flood protection, canals, and dams. Infrastructure has been a stalwart enabler of economic activity and has a powerful social presence. Despite agreement on the importance of infrastructure, economists, business leaders, planners, and politicians debate who should be responsible for infrastructure to meet economic and social goals, and who ultimately should own infrastructure. Both the public and the private sectors are active in the development and management of infrastructure, and ownership of infrastructure has gone through cycles of domination by one sector, and then the other. Who should own infrastructure

DOI: 10.1201/9781003396949-3

is a debate spanning centuries. There are periods in recent history when the private sector was the lead provider, followed by public takeover, then back to privatization, and now venturing into what are heralded as new partnership arrangements.

Infrastructure is a term that generally refers to tangible *things*—things that can be seen and touched; things that have life spans measured in decades, if not centuries. Today, infrastructure refers to both "social" and "economic" assets that deliver benefits to both the private and public sectors and may be privately or publicly owned, or a hybrid arrangement. Despite a widely held belief that infrastructure is the responsibility of government, the largest component of infrastructure is owned by the private sector and includes systems that deliver services that are essential to the growth and prosperity of individual businesses. Some of this infrastructure is part of the internal workings of private corporations in support of their operations, the costs of which are buried in corporate balance sheets and not easily identified.

Infrastructure assets are key components of socio-technical systems such as air, road, and water transportation; energy, water supply, and waste removal; health care and education; defense; and a wide range of production activities (Gil and Beckman, 2009). Only recently has infrastructure included a broad array of public and institutional buildings under the umbrella of social infrastructure. This list includes schools, hospitals, government buildings, libraries, clinics, jails, sports halls and arenas, and cultural facilities, previously referred to as public works, or real property assets. These are typically standalone structures that support the delivery of public services. The term infrastructure now extends to institutional housing such as student accommodation, social housing, facilities for the elderly, day cares, and even military housing. It includes new forms of economic infrastructure that are part of the invisible infrastructure critical to modern-day communications, comprised of satellites circling the globe, fiber-optic cable networks crisscrossing the oceans, and a myriad of land-based support structures. Beyond proprietary infrastructure networks such as oil or gas pipelines, there is the expectation that everyone is entitled to the services that infrastructure provides, at reasonable costs, and in a manner that ensures social equity and protects the public welfare. All of this makes defining infrastructure a challenging exercise.

The following quotations represent two definitions of infrastructure by two different economists.

From Kevin Kliesen and Douglas Smith (2009):

The nation's infrastructure can be thought of as its tangible capital stock (income-earning assets), whether owned by private companies or the government. This can include everything from the Toyota manufacturing plant in Indiana to the Fedex and UPS warehousing and distribution facilities in Memphis and Louisville, respectively. However, to most people, infrastructure is the nation's streets, highways, bridges, and other structures that are typically owned and operated by government.

From Edward Gramlich (1994):

> There are many possible definitions of infrastructure capital. The definition that makes the most sense from an economics standpoint consists of large capital-intensive natural monopolies such as highways, other transportation facilities, water and sewer lines, and communications systems. Most of these systems are owned publicly in the United States, but some are owned privately. An alternative version that focuses on ownership includes just the tangible capital stock owned by the public sector... Most econometric studies of the infrastructure problem have used the narrow public sector ownership version of infrastructure capital as their independent variable. This is in large part because it is very hard to measure anything else.

A key issue for supporting a definition of infrastructure is deciding whether to categorize assets by type, ownership, or function. There is a broad range of definitions of infrastructure and a marked contrast in definitions between public and private perspectives. The public side is prone to taking a limited stance in describing infrastructure. The OECD implies a common definition in their report *Infrastructure Financing Instruments and Incentives 2015*, when they state, "Traditionally, infrastructure investments have been financed with public funds. Governments were the main actor in this field, given the inherent public good nature of infrastructure and the positive externalities often generated by such facilities" (OECD, 2015).

Henry Cisneros, former Secretary of Housing and Urban Development (HUD) in the U.S., defined infrastructure capital as the structures and equipment that comprise "the basic systems that bridge distance and bring productive inputs together" (Cisneros, 2010). These definitions highlight the reference to serving the public interest, whether it be a dam, interstate highway, school, or hospital. They are also limited in scope, at least from a private perspective.

Economic studies often use all government-owned fixed assets as a measure of the infrastructure universe, but this definition includes public assets that many people still do not regard as infrastructure, such as schools or hospitals, and excludes private assets, such as private utilities or mobile phone towers. Most econometric studies of infrastructure use the narrow public sector ownership version of infrastructure capital as their independent variable. This is in large part because it is so hard to measure anything else. It is difficult to measure privately owned infrastructure, and even more difficult to distinguish private capital invested in infrastructure from private capital invested in other things. Compounding the problem is the difficulty of accounting for investment spending from consumption spending and how to deal with depreciation. With neither of these two types of spending is accounted for, it is difficult to arrive at any approximation of privately held infrastructure capital stocks.

Private investors are usually vague or ambiguous in describing what constitutes infrastructure, and with good reason. For example, they refer to "Permanent assets that a society requires to facilitate the orderly operation of an economy[1]". They may not infer a direct public good beyond the collective benefits of strengthening the economy and improving quality of life. However, there is common ground among economists on both sides of the argument in admitting that defining the boundaries of infrastructure is not a precise science and an exercise prone to subjective analysis (Bennett, 2020).

There is also an historical dimension to this discussion. Privatization of infrastructure may seem to some like a recent phenomenon. In fact, the debate over who should own infrastructure, private versus public ownership, is centuries old. In the U.S. in the 18th century, private canals and turnpikes were common and private transit systems were ubiquitous in 19th-century cities. Railroads, ships, and barges connecting cities, and linking sources of raw materials to points of production, and then to consumer markets, were private, and many remain so. The provision and distribution of energy is a mix of public and private. Despite these examples, when looking back at infrastructure in the U.S., broadly speaking, the public sector has built and managed infrastructure when the scale of investment was thought to be too large for private investors; the infrastructure generated positive externalities including health benefits, had nation-building benefits; or counter-recessionary macroeconomic benefits that would not be considered by private investors. The public sector also retained ownership where infrastructure could potentially be exploited by monopolistic private owners to the detriment of the public interest (Glaeser and Poterba, 2021).

Before George Washington became President of the United States, he was in the infrastructure business. He served as President of the Patowmack Canal Company with the intent to build a link between the eastern seaboard and the western waterways (Glaeser and Poterba, 2021). The original project faced engineering and financing difficulties and was moved farther north to be known as the publicly funded Erie Canal. New York Governor DeWitt Clinton, aware of the difficulties of securing enough private funding to create a massive infrastructure project, established the Erie Canal Commission which used public funds and public borrowing power to link the Hudson River to the Great Lakes. This is a 363-mile canal running from Albany to Buffalo, New York.

The Commission was an early example of an independent public entity overseeing an infrastructure project that relied on public financing[2]. The most famous 19th-century canals, such the Erie, the Erie and Ohio, and the Illinois and Michigan, were funded by states, not the federal government. The Erie Canal was enormously successful and user fees quickly covered the capital costs. The public sector was involved largely because capital markets were underdeveloped at the time and the public sector was the only plausible source of so much funding. These early schemes gave rise to the establishment of various public authorities at the state level with the power to design, build, and operate infrastructure such as bridges, toll roads, power plants, and energy distribution grids. Many still exist today, one of the best known being the Port Authority of New York and New Jersey, established in 1921.

Cutler and Miller (2005) document a strong link between public borrowing capacity and the construction of urban water and sewerage infrastructure during the late 19th century. U.S. cities and towns were spending as much on water at the start of the 20th century as the Federal government was spending on everything except the Post Office and the Army. The ability of cities and towns to borrow large sums in the financial markets enabled massive local infrastructure investments. This spending arrangement continues in the U.S. today.

The American story contrasts with pre-1800 English canal-building which involved smaller, flatter distances and private funds. For example, the original Mersey and Irwell navigation linking Manchester and the Irish sea was funded and built privately in 1734[3]. When Great Britain dug the much larger Caledonian Canal in 1804, public funding was used, but by the end of the 19th century, financial markets were sufficiently well developed so that the Manchester Ship Canal was a private enterprise.

Any discussion on public versus private involvement is further complicated by the rapid rise of digital infrastructure, which is particularly challenging to define as it typically incorporates both tangible assets such as cell towers and intangible assets such as software or other forms of intellectual property. The deployment and adoption of the commercial Internet in the 1990s brought about a major restructuring of digital infrastructure (Greenstein, 2021). Investment in digital infrastructure rose from about $25 billion annually in the 1980s to almost $250 billion in 2017 (Bennett et al, 2020). The sharp increase in digital infrastructure since the 1990s was driven by increases in investment in private communications equipment, as well as investments in software, the Internet, and computer hardware across almost all industrial sectors.

Today, digital infrastructure supports a range of innovative businesses in the sharing economy, social media, mobile information services, electronic retailing, ad-supported media, robotics, and artificial intelligence. These activities were much smaller, or nonexistent, in the 1990s. Their ascendancy and widespread market penetration changed dramatically in just a few decades. Digital services continue to grow and take on more importance in driving GDP growth. Because much of new investment represents evolving technologies, and because national accounts data may not be sufficiently granular to separately identify assets of interest, it is difficult to analyze these new sectors. Deciding what portion of a specific asset to allocate to digital infrastructure is challenging. For example, the equipment and software providing wireline and wireless access to the Internet could, in principle, be counted as part of cloud computing infrastructure and therefore included in a measure of digital infrastructure. However, these assets are also used for other purposes and dividing up these assets and sorting out these issues may be impossible (Glaeser and Poterba, 2021).

From an investment perspective, infrastructure is whatever matches the chosen infrastructure investment criteria established by a given investor. As investment criteria change in response to a highly competitive market for private infrastructure acquisitions, so does the definition of infrastructure. Investors continue to shift to a more inclusive version of infrastructure and continue to capture a broader array of tangible and non-tangible assets.

Notes

1 Caledon Capital Management. Slide presentation.
2 Sourced at https://en.wikipedia.org/wiki/Erie_Canal_Commission
3 Sources at https://en.wikipedia.org/wiki/Manchester_Ship_Canal

2 What we learn from history

Infrastructure has always played a key role in shaping and supporting patterns of urbanization and defining the modern world. Manhattan is identified by its unique grid of streets and utilities, the Commissioner Plan, with its many connecting bridges and tunnels. The character of Manhattan is defined by its raised and underground transit lines, and its dense, soaring verticality. Through the centuries, Manhattan has been one endless real estate play, capitalizing on the advancement of each new piece of technology to drive development upward in a land-constrained situation. Chicago is similar in many ways. It was built around the intersection of two rail lines adjacent to large body of water and a harbor, where the pressures of commercial activity necessitated new building forms. In the later part of the 19th century, Chicago gave birth to the high-rise office tower powered by electricity driving electric motors, water pumps, fans, and light bulbs, under the auspices of private infrastructure providers. There are two unique features of Chicago's infrastructure: the network of underground service roads that create an artificial ground level; and the elevated subway system serving downtown called the "Loop". These two infrastructure solutions, in combination, provide a separation of horizontal movement systems in Chicago's downtown seldom found in most urban centers.

Historically, advancements in infrastructure were driven by political ambitions and social needs and can be traced back centuries. In 1924, archeologists unearthed a planned city about 400 kilometers to the north-east of Karachi, Pakistan. The settlement was provided fresh water and sanitation through a clay pipe system (Jansen, 1989). Mahenjo-Davo flourished in the second half of the third millennium. Water came from more than 700 wells and supplied not only domestic demands, but also a system of private baths and a Great Bath for public use. Drains and sewers were carefully constructed to facilitate the removal of waste from each house.

Ancient Rome received more than one million gallons of fresh water daily from 11 aqueducts, the first built in 312 BC, and the remaining built over the next five centuries. Some still carry water, some 2,000 years later (Betz, 2020). The first sewers of ancient Rome are estimated to have been built between 800 and 735 BC. The sewage system did not really take off until the arrival of the

DOI: 10.1201/9781003396949-4

Cloaca Maxima, probably one of the best-known examples of sanitation from the ancient world. This "great sewer" was originally built to drain the low-lying land that ran through the Forum (ChemEurope, 2022). Over time the network of sewers that ran through the city expanded and most of them, including some drains, linked into the Cloaca Maxima, the contents of which were emptied into the Tiber River. In 33 BC, under the emperor Augustus, the Cloaca Maxima was enclosed, creating a large drainage tunnel.

The period of the classic Industrial Revolution, roughly 1760–1850, witnessed unprecedented growth and urbanization of the British population, and the transformation of the urban hierarchy as "new" industrial and manufacturing towns eclipsed older provincial and county towns. A consequence of this growth was the spread of cholera, a major global scourge in the 19th century, with frequent large-scale epidemics in European cities (Davenport, 2019). Private water companies proliferated in London in the early 19th century and by 1828 water was piped to over 150,000 mainly affluent households. Water supply was matched by a primitive sewer system, and this led to the installation of flush toilets in these affluent households which dumped untreated sewage into the rivers, the same rivers that supplied fresh water (Davenport, 2019). A noted English physician, Dr. John Snow, conducted pioneering investigations on cholera epidemics in England, and particularly in London in 1854, in which he demonstrated that contaminated water was the key source of the epidemics (Tulchinsky, 2018). His research indicated the difficulties that cities faced in evaluating the best means to provide infrastructure to ensure clean water and effective sanitation and waste disposal. A problem that still exists.

In 1878, Edison began working on a system of electrical illumination, something he hoped could compete with gas- and oil-based lighting. He began by tackling the problem of creating a long-lasting incandescent lamp, something that would be needed for indoor use. Edison did not invent the light bulb[1] but his use of this invention led to the founding of the Edison Electric Illuminating Company in New York City that installed 24 kilometers of wire illuminating 400 streetlights in a 1.5 square kilometer area of downtown New York[2]. His company became General Electric and his innovations in electrical illumination became a critical component of urban infrastructure.

As urban settlements expanded, people, products, and services had to be transported both within cities, and between settlements and the hinterland. Infrastructure had to be radically scaled up to support growing population densities. The essential building blocks of the industrial city included transportation, sanitation, fresh water, sources of energy, and a constant supply of food. Innovations led to disruptive technologies such as the steam engine, and advancements in new forms of infrastructure such as replacement of water wheels by coal-driven boilers. New forms of infrastructure defined the Industrial Revolution. In 1763, James Watt, a Scottish engineer, set out to improve Thomas Newcombe's simple piston engine used to pump seeping water out of coal mines. Watt unintentionally gave rise to the steam locomotive, and this opened a whole new era of infrastructure stretching across the globe (Rosen, 2010). Infrastructure was a strategic part of the colonization initiatives launched by many European countries, as evidenced by expansion of railroads as a crucial part of Britain's strategy to conquer and rule over India with only a small expeditionary force.

Largely forgotten are some of the more ingenious forms of infrastructure that accommodated city building in the past few centuries. For example, the installation of underground pneumatic mail networks, possibly a precursor of Elon Musk's recent Hyperloop prototype in Los Angeles. In 1887, New York City had approximately 45 kilometers of underground pneumatic tube linking 27 post offices across Manhattan[3]. It remained operational until 1952. In the 1870s, Paris installed a system of pneumatically synchronized clocks through a network of pressurized pipes below the street that served both homes and businesses[4]. But it was the abundance of coal, combined with new manufacturing technologies coming out of the Industrial Revolution, that dominated many of the infrastructure networks that we recognize today. The relatively short-term benefits of coal-based solutions led to an extraordinary increase in the number of wealthy individuals, and incremental improvements in quality of urban life in both Europe and North America for a rising middle class. The long-term legacy of fossil fuels is the environmental degradation that plaques the planet today.

This legacy had widespread consequences, severely damaging the health of city inhabitants. The Great Smog of London in December of 1952 is estimated to have caused between 10,000 and 12,000 deaths and over 100,000 severe illnesses[5]. Even today, the World Health Organization reports in their September 22, 2021, Newsroom—Ambient (outdoor) air pollution in both cities and rural areas was estimated to cause 4.2 million premature deaths worldwide per year in 2016; this mortality is due to exposure to fine particulate matter of 2.5 microns or less in diameter (PM2.5), which cause cardiovascular and respiratory disease, and cancers[6]. These environmental externalities are still not factored into the cost equation in most countries.

During a remarkable three-month period at the end of 1879, Thomas Edison installed the first practical electrical bulb, Kent Benz invented a workable internal-combustion engine, and a British-American inventor named David Edward Hughes transmitted a wireless signal over a few hundred meters. The introduction of new mass technology in the late 19th and early 20th century—telegraph, railway, electrification, radio, telephone, television, automobile, air travel—was accompanied by a utopian vision of the future and a host of ambitious promises, most of which were unfulfilled (Gordon, 2016). The naïveté of inventors in the 19th century was excusable, as no one could predict the full breadth of consequences from their revolutionary and fledgling innovations. However well intentioned, technical innovations which drove new infrastructure initiatives did not ease the yoke of heavy labor and promised long-term improvements in productivity.

In the U.S., total factor productivity (TFP), measured as average annual growth rate, in the period 1870–1900 was ~1.5–2.0 percent, rising to ~3.0 percent in 1930, and falling back to ~2.5 percent in 1940. It fell further in the period 1973 to 1990, averaging just ~1.0 percent and rose slightly in 2000 to ~1.5 percent (Shackleton, 2013). Productivity in the 20th century underlines the extent to which long-term TFP growth and economic growth in general have been influenced by the development of energy and transportation infrastructure suited to the expansion of suburbs. To the extent that policies that address the potential problems associated with climate change will require adjustments to patterns of land use, energy production and consumption, and transportation, those policies could have substantial but highly uncertain effects on TFP growth.

Infrastructure as we know it today is built below, at grade, and above the Earth's surface by both the public and private sectors. In 1863, the Metropolitan Railway Company, a private firm, built the world's first underground railway from Paddington Station to Farrington Street to the west and hence the term "Metro" became synonymous with subways. Despite many successful private initiatives, it is still undecided whether infrastructure that addresses critical societal needs should be public or private. Tokyo, for example, supports both a private and public transit system, both of which have lines extending across the city. Tokyo Metro, formerly the Teito Rapid Transit Authority, was privatized in 2002 and is now a joint-stock company owned by the Government of Japan and the Tokyo Metropolitan Government. Toei Subway, the Tokyo Metropolitan Bureau of Transportation, is an arm of the Tokyo Metropolitan Government[7]. The two subway systems operate on totally separate networks, and both are ranked among the busiest subway systems in the world.

It is difficult to examine the business models that underpinned infrastructure in the 18th and 19th centuries without references to historical precedents. For Europe, this was a period of nation building and a race to build global empires. Most notable were the British, who sought to extend their Empire through infrastructure networks extending to the far corners of the globe. The grandeur of this effort reflected ideological preferences to engage and enhance the role of the private sector, serving the interests of wealthy merchants, international investors, and powerful businessmen. It was loosely governed through powerful political appointments, with private chartered companies given unfettered reign over vast territories. The British Empire professed themselves as a "force for good" in the world to justify hostile occupation of foreign territories (Sartori, 2006). This rationalization provided wide license to quell resistance by force and introduced a range of approaches to building infrastructure across the Empire. The goal was to exploit resources and get these back to the homeland to continue to fund imperial ambitions. The result was a system of "command and control" that was always tenuous and chaotic, susceptible to a myriad of local pressures and problems, fraught with conflicts of interest, and the constant reshuffling of plans and priorities (Darwin, 2013). But it flowed copious amounts of cash and treasures back to the homeland and supported the opulence of the Victorian era.

One can trace this system of "command and control" back to The Ordinance of Staple passed by the Great Council in England in 1353[8]. This was the English way of flexing the government muscle, controlling the international movement of commodities, and taxing private enterprise. The Ordinance effected major ports in England, Wales, and Ireland and designated only those ports where specific goods could be exported or imported, with custom taxes imposed. A further intrusion to control private enterprises was the Statute of Monopolies, passed in 1623 by the English Parliament, that gave certain industries monopolies through a system of patent rights granted by government[9]. Around the same time, England launched several joint-stock companies to promote and control trade over large regions that were part of its colonization ambitions[10]. The East India Company, founded in 1600, was a joint-stock company that controlled large parts of the Indian subcontinent and rose to account for half the world's trade in the mid-1700s and early 1800s. The company eventually came to rule large areas of India, exercising military power and assuming administrative functions. The Hudson Bay Company, also a joint-stock company, received its royal charter in 1670. At its peak, the company controlled the fur trade throughout much of British-controlled North America. There was also the Levant Company or Turkish Company chartered in 1606, the Royal African Company charted in 1667, and the British South Africa Company chartered in 1889, all modeled on the East India Company.

This was the foundation of a system of commercial exclusion and the granting of monopolies intended to entice the private sector to participate in large-scale ventures. To succeed, these ventures required extensive private investments in infrastructure. To "modernize" India, the British built a vast railway network, and ports for their steamships, both of which are still in use day. This was accompanied by investments in irrigation, improvements in public health delivery, sanitation, policing, and a strong military presence. All were paid for through an oppressive tax collection system that crippled the Indian economy for centuries. The business model relied on the English Parliament engaging with private partners, notably hereditary nobles, wealthy local landowners, and a burgeoning merchant class.

If the British promoted monopolistic models to engage the private sector, it was the French who promoted the granting of concession rights with much tighter controls to secure the public interest. Dating back to the 17th century, many canals and bridges were built throughout France via the granting of concessions by government to private firms. Since the 19th century France has turned to private companies for a broad cross section of public services including water, sewage, refuse collection and treatment, urban transportation, mass housing and facilities for culture, sports, and social affairs (Hodge, 2015). The French did not view these arrangements as "sharing partnership", but as a form of engaging the private sector in the tradition of "government regulation of business", in which the public sector retained a significant degree of control and influence.

Private provision of a large segment of public infrastructure has been the norm throughout history. This began to change in the 19th century and continued throughout the 20th century, with the interrelationship between government and business evolving to accommodate new infrastructure initiatives. This shift was fueled by the apparatus of welfare economics which justified increased state intervention into market behavior and responding to the increasing disparity in the concentration of market power arising from economies of scale (Hodge, 2015). The main approaches to the delivery of infrastructure began to shift from direct intervention through public agencies and authorities contracting out for construction services, to the establishment of public regulatory bodies.

In Germany, significant public ownership of shares in private companies was deemed necessary to shape strategic infrastructure decisions. The French had their own unique approach through *the société economie mixte* (SEM) dating back to 1926, when local authorities were granted general powers to run local services, either directly or through financial participation with private firms[11]. Through SEMs the public sector maintained a significant degree of control and influence, and this promoted the early use of public–private partnerships in France.

Infrastructure projects under the control of the public sector are still largely carried out by single-purpose, centralized, supply-oriented utilities that operate in silos and capitalize on economies of scale. They access government funding, inexpensive resources, and public rights-of-way. Their interests are tied up in monopoly concessions. Many of these concessions were once granted by government to special interest groups and wealthy individuals, and these opaque silos have become part of the problem. Unfortunately, these practices are still integral to traditional professional training and institutional structures that govern infrastructure networks and perpetuate self-interest organizations.

Notes

1 Who invented the light bulb is a contentious topic. Humphry Davy demonstrated the first incandescent light to the Royal Institute in Great Britain, using a bank of batteries and two charcoal rods. Arc lamps provided many cities with their first electric streetlights. https://interestingengineering.com/science/who-actually-invented-the-incandescent-light-bulb
2 Sourced at https://en.wikipedia.org/wiki/Thomas_Edison
3 Sourced at https://untappedcities.com/2021/01/21/pneumatic-tube-mail-nyc/
4 Sourced at http://www.douglas-self.com/MUSEUM/COMMS/airclock/airclock.htm
5 "The Great Smog of 1952". metoffice.gov.uk. Archived from the original on 3 September 2014. Retrieved from Wikipedia 14 February 2022, website: https://en.wikipedia.org/wiki/Great_Smog_of_London
6 World Health Organization Reports September 22, 2021 Newsroom. Retrieved from www.who.int/teams/environment-climate-change-and-health/air-quality-and-health/ambient-air-pollution#:~:text=Ambient%20air%20pollution%20accounts%20for,quality%20levels%20exceed%20WHO%20limits

7 Sourced at https://en.wikipedia.org/wiki/Transport_in_Greater_Tokyo

8 Sourced at https://en.wikipedia.org/wiki/Statute_of_the_Staple

9 Sourced at https://en.wikipedia.org/wiki/Statute_of_Monopolies

10 A joint-stock company consisted of investors who pooled resources to fund an enterprise and, if it was successful, shared the profits. Using such an arrangement to fund colonial ventures proved to be attractive both to the Crown and to investors.

11 Sourced at https://fr.wikipedia.org/wiki/Soci%C3%A9t%C3%A9_d%27%C3%A9conomie_mixte

3   Who owns infrastructure

Does it really matter who owns infrastructure? The answer is a resounding yes from all sides of the ownership debate, but for a variety of different reasons. Who should own infrastructure, and why, are two of the most contentious topics when dealing with infrastructure. Responses to these two questions directly affect the level and type of private engagement and funding of infrastructure. Responses will vary depending on who you are talking to, where, and when the conversation occurred. Private sector business is grounded in forms of ownership and the ultimate level of control that privatization implies. The public response is often fraught with misunderstanding on what ownership means, what infrastructure they control despite not having ownership, and not always knowing who owns what infrastructure that users rely upon.

Contrary to popular belief most infrastructure is privately owned, often buried in corporate holdings. Ownership is also affected by how infrastructure is defined and regulated. On the public side there is the misconception that it is federal or national governments that own most of the infrastructure. This is misleading. In developed countries, most public infrastructure is owned or controlled at the state or municipal level. In comparison, ownership by national governments is relatively small and has been declining for decades. This is not the case in most emerging economies.

In "A brief history of infrastructure in Canada, 1870–2015", Professor Herb Emery at the University of Calgary points out that since 1961, over 75 percent of infrastructure capital stock has been provincial and local infrastructure. By 2002 that number rose to over 90 percent; the local share rising from 30 to 50 percent, provincial declining from 45 to 42 percent, and federal share declining from 25 to 7 percent (Emery, 2015). This reflects the fact that over this period much of economic growth has been urban growth. Professor Emery also points out that, "History tells us that focusing on publicly owned infrastructure may be misleading" and that "a lot of infrastructure is developed by private owners".

The Federal share of public infrastructure in Canada is small (LOPRESPUB, 2016). In 2013, the Federal government owned less than 2 percent of roads, bridges, public transit, water, wastewater, culture, recreation, and

DOI: 10.1201/9781003396949-5

communications infrastructure in Canada. Provincial/territorial governments held 41 percent of this "core" public infrastructure and municipal governments the remaining 57 percent. Over the past 50 years, the Federal government has scaled back its holdings of road, water, and wastewater stock (Figure 3.1). This is consistent with its division of powers with provincial governments (Sections 91 and 92 of the *Constitution Act, 1867*). The Federal government further reduced its involvement in infrastructure by disposing of its ownership of air and rail freight services during the 1980s and 1990s. It retained ownership of the passenger rail service, which has been on life support ever since.

The situation in the U.S. is not dissimilar to that of Canada. In 2017, according to the Congressional Budget Office (2018), the U.S. Federal Government spent $98 billion on transportation and water infrastructure. State and local governments spent $342 billion (Glaeser and Poterba,2021). To find consistent data on infrastructure in the U.S, the best source is the National Income Accounts produced by the US Bureau of Economic Analysis (BEA). For the public and private sectors, the BEA calculates the net stock of fixed assets, which is a very broad but uniform measure of infrastructure.

In the U.S., infrastructure is characterized by highly decentralized ownership and funding. State and local governments and the private sector own 97 percent of the nation's non-defense infrastructure and collectively fund 94 percent of this infrastructure (Edwards, 2017). But which infrastructure are they referring to? Economists admit to the difficulty in establishing clear criteria to decide what assets should be considered infrastructure. One database that is sourced for hard and consistent data on infrastructure in the U.S. is the National Income Accounts produced by the Bureau of Economic Analysis

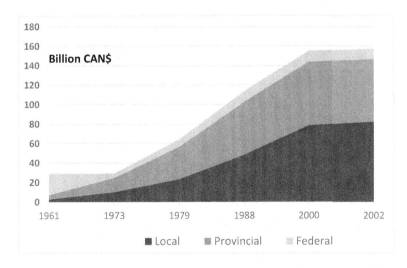

Figure 3.1 Total value of the stock of public infrastructure, Canada 1961–2002.

Source: Harchaoui et al. Statistics Canada. 2003. Net of linear depreciation.

(BEA) which calculates the net stock of fixed assets. This is a very broad but uniform measure of infrastructure[1]. Figure 3.2 shows that the private sector owns most of the nation's non-defense infrastructure. In 2015, private infrastructure assets of $40.7 trillion were four times larger than state and local assets of $10.1 trillion, and 27 times larger than federal assets of $1.5 trillion[2]. These figures raise the question of the efficacy of any national infrastructure plan or determining the most effective federal intervention strategies to promote infrastructure (Edwards, 2017).

A closer look at government-owned infrastructure assets in the U.S. indicates that the Federal government owns just 13 percent of the total, while state and local governments own 87 percent (Figure 3.3). The Federal government dominates in two areas: intellectual property and land conservation. However, the Federal government in the U.S. exercises a significant level of control over state, local, and private infrastructure through taxes and regulation, neither of which receive sufficient recognition for their efficiency and effectiveness.

Public versus private

Who should own or control infrastructure is a debate that has been centerstage throughout history. It is an argument fraught with ideological overtones, popular sentiment, political posturing, legacy issues, regulatory regimes, and the taxing structures applied by a nation, state, or city. A confusing blend of public and private ownership of infrastructure exists throughout the world, a blend that often defies a logical explanation. We have both private and public

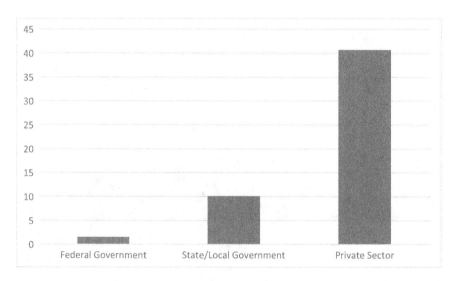

Figure 3.2 Net stock of fixed assets, 2015 (trillions of dollars).

Source: U.S Bureau of Economic Analysis, 2015. Excludes defense.

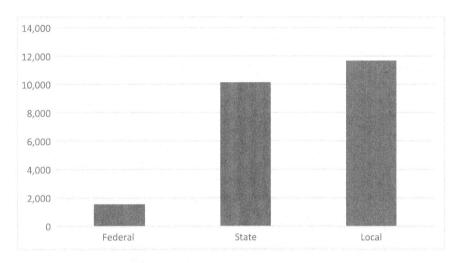

Figure 3.3 Net stock of government fixed assets, 2015 (billions of dollars).

Source: U.S. Bureau of Economic Analysis, 2015. Excludes defense.

transit systems, rail lines, airports, water systems, utilities, waste management, and energy companies, often side-by-side in the same city. Communications networks are largely private, as are many energy producers and distributors. There are public and private hospitals, schools, cultural centers, sports halls, arenas, and parking garages throughout most cities. This gives rise to ongoing debates over who should pay for infrastructure—governments or end users. Other critical areas of discourse include the efficacy of public monopolies, where the line between public and private ownership should be drawn, and how to maintain and operate these assets over long periods of time, in some cases over several lifetimes. Public regulatory regimes add another layer of complexity to the debate.

Central to most discussions about infrastructure are two defining issues: to what extent should some infrastructure be deemed an essential public service paid for by taxpayers; and should we even consider privatizing what are traditionally seen as public assets. A classic dilemma is the case for privately built and operated toll roads on a public right-of-way and under a concession agreement, versus unrestricted access to all major public thoroughfares, even if under ineffectual government ownership, as evidenced by the current infrastructure crisis in the U.S. A most recent example is the rejection by a Court in Pennsylvania of a PPP to repair and replace nine decrepit bridges across the state (Bentley, 2022). A State senator, and chairman of Pennsylvania's Senate Transportation Committee, commented that "Today's decision is a win for all Pennsylvanians. A win for all those who stood with us fighting this oppressive overreach". The court ruled in favor of the Appellant and against the State's

Department of Transportation and its PPP Board. The judge found that this major PPP bridge initiative to be *"void ab initio"*, or without legal effect from inception. The ruling was a clear message that PPPs were not welcome in Pennsylvania.

There is a confusing array of infrastructure issues, many of which defy rational explanation as per the comments of the Pennsylvania senator. Approaches to providing infrastructure influence how we build cities and nations, fuel our prosperity, address threats and challenges to society, or respond to new public policies and procedures. When it comes to cities, explanations of our inter- and intra-city infrastructure become even more confusing. What lies behind this apparent confusion is a complex web of people, policies, politics, and processes, spread across long time periods.

This debate is highlighted when examining the level of services required by private residences housing citizens in urban centers. Each home requires electricity, heating and/or cooling, clean water, sewage and sanitation, police and fire protection, storm water management, and high-speed Internet access. To this list can be added parks and playgrounds, libraries, and public health clinics. Residents require access to road and highways, public transit, bicycle paths, and parking to facilitate getting to and from work, shopping, and recreation. For intra-city travel, residents require airports, bus and rail lines, and transportation terminals. Who decides what should be publicly versus privately operated? Who can offer the best quality services for the most reasonable cost? There is not a logical breakdown between public and private infrastructure services at the level of the household or within cities and regions.

In the past several decades public infrastructure has caught the attention of private investors and large institutional investors such as pension funds, although their total investments in infrastructure represent an insignificant fraction of what is needed. The pressures of urbanization, the migration of people from rural areas into cities, and the inability to address continued deterioration and obsolescence of existing infrastructure, are forcing an expanded conversation on the merits of new approaches to harness the involvement of the private sector to meet infrastructure challenges. Today, in Western nations, this is still a hot topic from political, social, and economic perspectives.

The origins of a shift to increasing privatization of infrastructure was reignited by the neoconservative movements in the UK and U.S. in the early 1980s. Prime Minister Margaret Thatcher introduced sweeping changes following her win in the UK general election of 1979. Her political philosophy and economic policies emphasized deregulation, particularly of the financial sector, flexible labor markets, the privatization of state-owned enterprises, and reducing the power and influence of trade unions. Privatization has been called "a crucial ingredient of Thatcherism" and her government went on to privatize utilities (gas, water, and electricity), airports, harbors, and railways, although she resisted the privatization of railways. This task she left to her successor, John Major, who implemented rail privatization in 1994[3]. The subsequent rapid decline of rail service in the UK became a rallying point for those opposed to privatization.

Fueling this ideological shift to privatization was the pragmatic realization that governments did not have the necessary resources to do the job in the face of growing expenditures for health care and the impacts of an aging population. Privatization and the use of private funds were seen as the best alternative to raising taxes or increasing the national debt. New arrangements were sought that offered a reasonable option to outright privatization and involved the public and the private sectors in contractual arrangements to design, develop, finance, operate, and manage new infrastructure, in lieu of governments handling these tasks internally. Government wanted to rid itself of two risks that drove up costs the public sector could not manage—construction risk and scheduling risk. This brought about the adoption of a new procurement model based on risk transfer and involving a complex set of relationships between people, policies, and processes that involved the suppliers of engineering and contracting services, consultants, manufacturers, financiers and investors, developers, and both existing and new infrastructure. Thus began the evolution of public–private partnerships or PPPs, building on the earlier platform of the power-purchase agreement devised in the energy sector back in the 1970s. With PPPs, a link was forged between infrastructure and private business that is still evolving.

An ever-changing pattern of ownership and control

One of the great challenges in dealing with infrastructure is that of uncertainty: matching durable, long-term assets whose value is predicated upon continuity of services, to addressing the demands of an ever-changing society and the vagaries of political masters. Examining the history of the inter-state highway network in the United States is a good example of the fluidity and discontinuity that befalls large-scale infrastructure systems.

The launch of the U.S. interstate highway system was hastened by the proliferation of privately owned cars and commercial vehicles in the early 20th century. This asset can be described as a continuous roadbed in a designated public right-of-way, and includes support structures such as bridges, tunnels, and on-and-off ramps. The origins of a national road system date back to the Federal Road Act of 1916 that provided matching Federal funds to individual states to build and improve highways. The Act was short-lived with the advent of World War I and was replaced by the Federal Aid Highways Act of 1921, again based on matching public funding sources[4]. This Act specifically targeted the construction of a national grid of interconnected primary highways. The Army was asked to provide a list of eligible roads, and this created the so-called Pershing Map, named after General John J. Pershing, who directed the effort[5]. This gave birth to the "superhighway", and hence to what became the inter-state highway system. But it took another Army general to make it happen.

President Dwight D. Eisenhower championed the Federal Aid Highway Act of 1956. He was influenced by his position as Supreme Commander of the Allied Forces in Europe during World War II, where he experienced the advantages of Germany's autobahn network. He was concerned with the lack of key ground transportation routes in the U.S. to move military supplies and troops in case of emergencies or foreign invasion. This became known as the Defense Highway Act of 1956. Prior to the Federal Highway Act and the establishment of the Highway Trust Fund in the same year, roads were financed directly from the General Fund of the U.S. Department of Treasury. The highway system that these two Acts launched was proclaimed completed in 1992, despite all links not being in place due to strong local opposition to inner-city links through dense urban areas such as downtown Boston and San Francisco.

As far as ownership and financing, highways and rights-of-way in the system are owned by the state in which they are built. Approximately 70 percent of construction and maintenance costs are paid through direct and indirect user fees, primarily fuel taxes collected by federal, state, and local governments. To a much lesser extent, costs are paid for by tolls collected on toll highways and bridges. The remaining costs are borne by general revenue receipts, bond issues, property taxes, and other forms of taxation. The overwhelming source of Federal funding (93.5 percent in 2007) is derived from fuel tax. At the state level user, fees in the form of tolls represent about 60 percent of their total contribution. Local governments are generally averse to user fees to fund their contribution and typically cover their portion of the cost from annual capital budgets.

With the success came heavy usage and rapid wear and tear which soon caught up with the interstate highway system and exposed an inadequate funding formula for maintenance and repair. The rapid growth of the post-War suburb expanded inter-city networks outwards, resulting in rapidly rising maintenance costs that left fewer funds for new intra-state highways. In 1983, Congress permitted funds in the Highway Trust Fund to be used for mass transit systems, placing further strain on diminishing financial resources for highways. In 2008, the Fund required an $8 billion transfer from general revenues to cover a deficit. This shortage was due to lower gas consumption because of the recession and higher gas prices.

Before 2008, highway tax revenue dedicated to the trust fund was sufficient to pay for outlays from the fund, but after 2008 that was no longer the case. Since 2008, Congress has sustained highway spending by transferring over $140 billion of general revenues to the fund, including $70 billion in the Fixing America's Surface Transportation Act in 2015 (Brookings, 2022). The Congressional Budget Office projects that, by 2030, outlays from the Highway Trust Fund will exceed trust fund reserves by a cumulative $134 billion for the highway account and by $54 billion for the mass transit account, even if expiring trust funds taxes are extended (Congressional Budget Office, 2020).

A lack of funds pushed states to increasingly turn to toll roads to meet maintenance and expansion demands. Although federal legislation originally banned the collection of tolls on interstate highways, with the exception of existing toll roads and bridges that were "grandfathered in", the Federal position changed in 2005. New legislation encouraged states to construct new interstate highways through "innovative financing" methods, easing restrictions on building new toll roads, and promoting public–private arrangements. However, the legislation still prohibited the installation of tolls on existing toll-free interstates without prior approval from Congress, to be considered on a case-by-case basis. Today, the U.S. interstate highway system is a confusing network of chargeable and non-chargeable routes, implemented on a seemingly ad hoc basis, and reflecting the influence of politicians at all three levels of government. What remained constant over the years is the accumulation of massive, deferred maintenance costs, a problem yet to be resolved.

Fast-forward to the possibility of electric vehicles (EVs) dominating over convention fuel-powered vehicles, plus the prospect of autonomous cars and commercial vehicles. The picture is transformed once again. A whole host of radically different infrastructure improvements will be required including charging stations and monitoring and data collection devices. Who will pay for what, and how? This will mark a radical shift in the use of the interstate highways and old funding models will come into question, primarily the reliance on gasoline taxes as a major source of revenue. The interstate highway example illustrates the level of complexity in defining infrastructure in terms of the characteristics of a hard or fixed asset, without acknowledging the shaping and reshaping of consumer demands over time, evolving technologies, and the fluidity of government mandates. Another element of the U.S. transportation system, railroads, is what John Cassidy refers to as a "parable of contemporary American capitalism" (Cassidy 2022).

A downside of privatization

Cassidy claims that privately owned railroads, one of the oldest and far-flung industries in the U.S., are "a story of deregulation, consolidation, downsizing, underinvestment and intensification". He cites this as an example of prioritizing payments to wealthy stockholders over everything else, including serving the public interest. There are now just seven large Class 1 railroads in the U.S. compared to 33 in 1980, and between them they control more than 80 percent of the freight market. Martin J. Oberman, current chair of the Surface Transportation Board, noted in a speech last year that between 2011 and 2021, the big railroads spent 190 billion dollars on dividends and stock buybacks, far more than the 138 billion dollars they spent on capital improvements in the same period (Cassidy, 2022). A similar story is emerging with privately owned water systems in England. Six water companies are under investigation for potentially illegal activities as pressure grows on the industry to put more money into replacing and restoring crumbling infrastructure to protect both the environment and public health (Leach, 2022).

New ownership structures

Purchase power agreements (PPAs) illustrate a model of delivering infrastructure based on the principles of project finance. PPAs deliver essential services that combine public and private ownership in a contractual arrangement that has evolved and is still widely in use today. Off-take contracts offer exclusive rights to one or more wholesalers to use a specified portion of output capacity for a set period of time under a PPA. In exchange, the beneficiary company pays fixed period sums to the producer (a private special purpose vehicle). On the input side, based on an output contact, the producer pays fixed period sums for the required inputs such a natural gas to run generating turbines (a fuel supply agreement, or FSA). This contract model for distributing electrical power was first used on a wide scale in the United States and various European countries to develop private power plants (Gatti, 2008). The PPA contract became the precursor for the development of PPPs, a form of public sector procurement. Through PPAs, large sums of long-term private capital are raised for large capital-intensive infrastructure projects, particularly in the energy and mining sectors.

The use of project finance upon which PPAs are founded dates back to the 18th and 19th centuries. However, the popularity of PPAs in more recent decades can be traced back to their use in the U.S. to address the energy crisis in the 1970s. The crisis precipitated passage of the Public Utilities Regulatory Policies Act (PURPA) and the Fuel Use Act by Congress in 1978 (Bloom, 2015). The first Act reduced barriers to promoting alternative energy developments and the second Act promoted cogeneration gas-driven power plants. These two acts defined the respective roles of the public and private sectors as active market participants as an alternative to traditional public utility arrangements. PPAs have gradually evolved into other types of contract models, an example being *tolling* (Gatti, 2008). Tolling enables the energy producer, usually a special purpose vehicle (SPV), or an independent power producer (IPP), to generate sufficient cash flow to repay the initial investment. It was used in countries such as the U.S. and the UK, and later in EU countries that liberalized domestic markets for electricity and gas supplies.

The PPA (Figure 3.4) provides for a privately owned project company (special purpose vehicle) to construct a power station with specified technical characteristics. Construction carries a specified delivery date and then it is operated on an agreed-upon basis (Yescombe and Farquharson, 2018). The power is sold under a long-term tariff agreement to the power purchaser who may be a public-sector transmission and distribution company, a local distribution company, or a direct end-user of the power. The tariff is usually paid monthly by the power purchaser to the project company and has two components: a fixed availability charge and an energy charge (or variable charge) which varies with the usage of the plant. A concession agreement is entered into between a public-sector entity and the project company under which the project is constructed to provide a service to a public-sector entity, or directly to the public.

PPAs and PPPs look remarkably similar in that they are both are based on project finance and center on a privately owned SPV. Key features include financing of capital costs through a combination of shareholder equity and

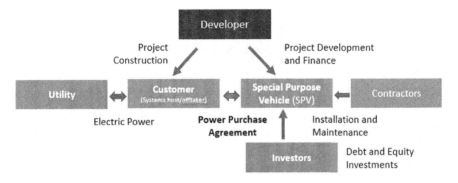

Figure 3.4 Conceptual framework for a power purchase agreement (PPA).

Source: Better Buildings: U.S. Department of Energy. Power purchase agreement.

project finance debt (Yescombe and Farquharson , 2018). Through a series of subcontracts, a contractor agrees to:

- Construct a facility under a design/build (DB) contract, based on detailed specifications and at a fixed price and delivery schedule.
- A "soft" facilities management (FM) contract with a service provider that covers services such as security, cleaning, and other maintenance services.
- A "hard" FM contract with a maintenance company or the D/B contractor provides for building maintenance services.
- A project agreement that defines the involvement and relationship with the public authority or entity.
- A cash flow waterfall that covers operating cost payments, debt service costs, and then a distribution to the investors.

Building on the structure of PPAs, the growth in PPPs (Figure 3.5) over the last few decades has been driven by several factors including the public-sector reform movement known as "New Public Management" (NPM) that arose in the 1980s (Charbonneau, 2012). A significant appeal was the ability of PPPs to circumvent public-sector budgets through off-balance sheet accounting for incurred financial liabilities. This feature alone was sufficient in many cases to justify a decision to use a PPP as it circumvented fiscal constraints and, through a concession agreement, maneuvered around the ownership debate. Governments with limited access to capital could borrow privately off-balance sheet and not increase reported debt numbers, thereby exhibiting fiscal prudence.

Through a concession agreement, a PPP allowed government to maintain legal ownership of an asset but assign beneficial ownership to a private entity for an extended period of time. Those in favor of PPPs claim a number of

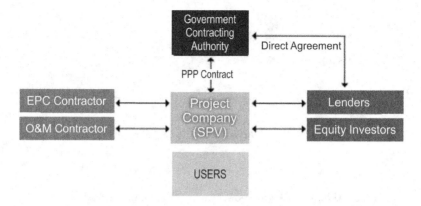

Figure 3.5 Typical public–private partnership (PPP) project structure.

Source: World Bank. Public-Private Partnership Legal Resource Center, June 2022.

advantageous features: permitting a public authority to remain politically accountable for an essential public service; avoiding the specter of outright privatization; preserving the right to limit competition through a government's monopolistic control; retaining the long-term ownership of the assets in public hands; and providing a range of public service from a private entity at pre-determined costs.

Detractors of the PPP model point to the higher costs involved in the "slight-of-hand" accounting that distorts total public debt and indentures public assets. They also question the cost of retained risks including political risk, the ability to enforce contracts over long time periods that can never be deemed "whole", and the accuracy of the value-for-money proposition that governments use to justify their decision to proceed with PPPs. Seldom do public bodies factor in the possibility of failure, the need to renegotiate agreements, or even sale of the agreement by the private party during the term of the concession. There is always the hidden question as to who is at risk when a PPP project defaults. The owner of course. Most arguments against PPPs assume that governments have real choices in the matter, which often they do not have. It is important to note that in the absence of PPPs, investment in a broad range of infrastructure projects would never have materialized, particularly in the case of social infrastructure, and for countries struggling to attract financing and funding to address infrastructure needs.

The initial success of infrastructure delivery models, such as PPAs and PPPs, has not tempered the debate on ownership of infrastructure. These are complex delivery models, intertwining public and private interests, and involving large amounts of capital. They appear to favor large private-sector organizations, private equity that is relatively expensive, and long-term commitments. PPPs never make it clear to the taxpayer who owns or controls what, and why. The fact that public costs are buried in government accounts compounds the problem of determining who are the winners and losers in these arrangements. Critics claim that PPPs guarantee private profit at public expense. This places the debate right back into the political arena where the arguments are neither black nor white. Playing the ownership card is in the interests of politicians who must judge the sentiment of taxpayers and their political base. The ownership debate can be sufficiently rancorous to defer a decision or opt to do nothing.

Notes

1 Bureau of Economic Analysis capital stock data.
2 The net stock of defense assets in 2015 was worth 1.7 trillion.
3 The ongoing controversy surrounding privatization of railroads in the UK has proven to be a most contentious issue and held up as an example by critics of why privatization of infrastructure is not in the public interest. In December 2017, Lord Adonis resigned his position as head of the National Infrastructure Commission. Among his primary reasons was the taxpayer-funded "bailout" of Stagecoach and Virgin—the private operators of the East Coast rail line—after Virgin Trains East Coast, a partnership between Virgin and Stage Coach, had been allowed to walk from an agreement by which they had agreed to pay Government £3.3 billion to run the service until 2023.
4 Sourced at https://en.wikipedia.org/wiki/Federal_Aid_Highway_Act_of_1921
5 https://en.wikipedia.org/wiki/Pershing_Map

4   Addressing the infrastructure gap

The need for new public infrastructure and maintenance of existing infrastructure is not being met and this dilemma is universal. The bottleneck is not the availability of funds, but the inability to increase the flow of funds to address infrastructure needs from sources other than public coffers. The paradox is referred to as the "infrastructure gap"—the ever-widening gap between what the world currently invests in infrastructure and what it needs to invest to meet growth and climate-related needs. The infrastructure gap must be seen in the context of a multitrillion dollar global market for "hard" assets, historically dominated by the developed countries, with an increasing awareness of the growing needs in the emerging markets. This is evident in the major disconnect between institutional investors (insurers, pension funds, and sovereign wealth funds) seeking investment opportunities and infrastructure projects requiring capital. For the past two decades there have been increasing amounts of private capital in pursuit of investment opportunities in infrastructure, however, this increasing reserve of capital faces diminishing opportunities to invest in infrastructure right across the globe.

The issue for investors is not finding money but spending money. This has precipitated a funding gap in four key infrastructure sectors and the magnitude of the gap in each sector illustrates the severity of the funding shortfall. The world invests approximately $2.5 trillion annually in transport, power, water, and telecom—whereas the need is for $3.3 trillion annually just to meet growth forecasts to 2030 ((Woetzel et al., 2017). The gap is the difference between $2.5 trillion that is funded and the $3.3 trillion that is needed. Why does this gap even exist when both the public and private sectors are aligned in their appreciation of the socioeconomic benefits associated with building infrastructure?

McKinsey estimates that infrastructure typically has a socioeconomic rate of return of around 20 percent (Woetzel et al., 2017). In other words, one dollar of infrastructure investment can raise GDP by 20 cents in the long run and some infrastructure investments, if well-chosen and well executed, can have benefit–cost ratios of up to 20:1. Estimating the magnitude of the gap can be misleading and disguises the complexity of the shortfall. The infrastructure gap varies considerably across the Western world, within countries,

DOI: 10.1201/9781003396949-6

and across sectors. In the West, the gap can represent a significant cost for repairing deferred maintenance of existing infrastructure, both social and economic infrastructure, while in the emerging economies the primary need is for new infrastructure to meet the rapid pace of urbanization. Closing the gap also has significant geo-political dimensions and the focus will increasingly shift to the emerging economies where need is the greatest. Climate change will dramatically increase the size of the gap in future years.

Assigning numbers to the gap can take many forms. It may refer to percent of annual GDP, annual dollar amounts, or cumulative amounts over a given time span. Various organizations have mapped out estimates of the infrastructure gap, and no matter how measured, the numbers are staggering. According to the Global Infrastructure Hub (GIHub), a G20 initiative, global infrastructure investment needs to target $94 trillion between 2016 and 2040 versus the current trend, which is estimated to be $79 trillion, resulting in an investment shortfall of $15 trillion (GIHub, 2017). This is 19 percent higher than would be delivered under current trends and averages $3.7 trillion per year. Meeting the UN Sustainable Development Goals for universal access to drinking water, sanitation, and electricity by 2030 increases the global infrastructure need by a further $3.5 trillion by 2030. To meet this investment need, the world will need to increase the proportion of GDP it dedicates to infrastructure to the 3.5 to 3.8 percent range, compared to the 3.0 percent expected under current trends (Figure 4.1).

McKinsey in its report, Bridging Infrastructure Gaps, estimates average annual needs, 2017–2035 in US$ trillions, indicates a requirement to meet an annual need in this period of 4.1 percent of world GDP, equivalent to $69.4 trillion in aggregate spending in this period (Woetzel et al., 2017). The largest gaps are in power ($20.2 Tr) and roads ($18.0 Tr), representing 55 percent of the total aggregate need. Emerging economies account for some 60 percent of this need. If the current trajectory of underinvestment continues, the world will fall short by roughly 11 percent, or $350 billion a year.

The size of the gap triples if we consider the additional investment required to meet the UN's Sustainable Development Goals (SDGs). In contrast to growing needs, infrastructure investments have declined as a share of GDP in 11 of the G20 economies since the global financial crisis in 2008, despite glaring gaps and years of debate about the importance of shoring up foundational systems. Cutbacks in infrastructure spending have occurred in the European Union, the United States, Russia, and Mexico. In comparison, Canada, Turkey, and South Africa increased investment, but nothing compared to the increases in China.

One must separate out China in this discussion given their pace of investment over the last several decades in infrastructure compared to that in Western nations. In the period 1992 through 2013, China averaged an annual expenditure level on infrastructure of 8.6 percent of GDP. Much of this new infrastructure was in power and roads. Compare this to the United States and Canada, who each averaged around 2.5 percent of GDP in the same period, with the largest component being telecom. China spent more on economic

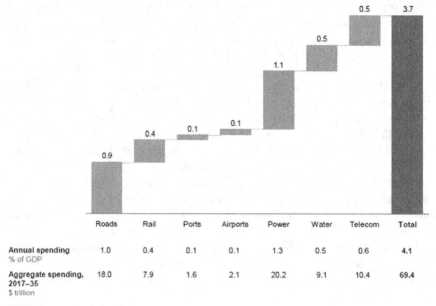

Average annual need, 2017–35

$ trillion, constant 2017 dollars

	Roads	Rail	Ports	Airports	Power	Water	Telecom	Total
Annual spending % of GDP	1.0	0.4	0.1	0.1	1.3	0.5	0.6	4.1
Aggregate spending, 2017–35 $ trillion	18.0	7.9	1.6	2.1	20.2	9.1	10.4	69.4

NOTE: Numbers may not sum due to rounding.

Figure 4.1 Estimated average annual need, 2017–2035.

infrastructure annually than North America and Western Europe combined. In 2013, China represented one-third of total global investment in infrastructure and there are no signs that the pace of investment slowed in the subsequent decade. However, given recent events in the country, there are recent signs that China's rate of spending on infrastructure may be declining.

Impact of the infrastructure gap

In 2007, the Minneapolis I-35 Bridge collapsed during rush hour, killing 13 and injuring 145 (Figure 4.2). In 1990, the federal government rated the bridge—and 75,000 others—as "structurally deficient"[1].

The impacts of the infrastructure gap are evident across the globe. The glaring examples are broken or collapsed bridges, tunnels now beyond repair that must now be replaced, constant power blackouts, flooding, deteriorating dams, tainted water supplies, lack of sanitation, and congested throughfares.

Figure 4.2 The Minneapolis I-35 Bridge collapse, 2007.

Source: Minneapolis Department of Transportation.

On the climate-related side is the continuing impact of Hurricane Katrina, a large and destructive Category 5 Atlantic hurricane that struck New Orleans, Louisiana, and caused over 1,800 fatalities and $125 billion in damage in late August 2005[2]. The severity of the damage was largely blamed on deteriorated flood protection infrastructure. In September 2022, Hurricane Ian hit the State of Florida and its Gulf Coast before making final landfall in South Carolina. Ian is the second-deadliest storm to strike the continental U.S. this century, behind Hurricane Katrina. At least 101 people are confirmed dead.

The infrastructure gap, presented as a cumulative total, tends to be interpreted as a problem for national governments, when in many countries this is not the case. Social and economic infrastructure in most countries is primarily the responsibility of municipal and regional governments who have very limited funding capacity beyond annual property tax receipts and some limited user fees. Cities are usually at the mercy of upper-tier governments who control legislation and collect the largest share of taxes. Canada is a good example.

The Canadian Federal government owns just 1.8 percent of the net stock of core public infrastructure, the provinces and territories own 41.4 percent, and municipal governments own 56.8 percent (PPP Council, 2016). The Federal government has the greatest capacity to raise funds through federal taxes, and other levies on trade and commerce, plus utilize deficit spending with access to the bond market with a AAA sovereign rating[3]. They can also offer incentives for investment by grants or tax measures. Provinces can also run deficits and issue bonds. The Province of Ontario, with the largest population in the country, is ranked as an "extremely strong" investment-grade borrower.

In comparison, the metropolitan area of Toronto, with a population of 6,255,000 (2021), and a growth rate of close to one percent annually, must meet its infrastructure needs almost exclusively from property tax receipts and the largesse of grant programs from the Federal and Provincial governments. The city must present a balance budget, cannot run deficits, and requires Provincial legislation and approval to introduce new sources of revenue. Closing the infrastructure gap at the municipal level is fraught with funding challenges.

The claim in the U.S. is that, after decades of allowing critical infrastructure to fall into serious disrepair, the country is about to embark on a new era of federal funding for infrastructure directed at state and local governments. The Bipartisan Infrastructure Law, signed into law in November 2021, includes $1.2 trillion in total spending (baseline and new program spending) of which $550 billion is new program spending. A total of $470 billion of this is directed to state and local governments through a competitive bidding formula and $180 billion to state and local governments through competitive grants (Dunn, 2022). Most of this spending is based on the use of U.S. tax receipts.

At a global level, Asia will dominate the global infrastructure market in the years ahead, as it does at present (Woetzel et al., 2017). Asia accounts for some 54 percent of global infrastructure investment needs to 2040, compared to 22 percent for the Americas, the next largest region. Indeed, just four countries account for more than half of global infrastructure investment requirements to 2040: China, the U.S., India, and Japan. China alone is estimated to account for 30 percent of global infrastructure needs. The infrastructure investment gap is proportionately largest for the Americas and Africa. Comparing forecasts of infrastructure need to what would be delivered under current trends results in estimated investment needs in the Americas at 47 percent greater than forecast investment under current trends. For Africa, the equivalent figure is 39 percent (Woetzel et al., 2017). While the latter offers considerable growth potential, the African infrastructure market remains small in absolute terms. The region accounts for just 6 percent of global infrastructure investment needs.

As a relatively mature infrastructure market, Europe tends to invest less in infrastructure as a proportion of GDP than regions which include more low- and middle-income countries (Woetzel et al., 2017). Overall, Europe invested 2.2 percent of GDP in infrastructure between 2007 and 2015. This is the second lowest proportion amongst the regions in the GIHub study, behind only the Americas. To deliver the infrastructure investment identified by forecasting current trends, Europe would need to maintain spending at a similar share of GDP between 2016 and 2040 or increase it slightly to 2.6 percent of GDP to meet the investment needs forecast.

Closing the gap requires five things:

- First, the political will to prioritize infrastructure needs. As with any public investment in large-scale projects, political support is imperative, and this will not be easy in the post-COVID era.
- Second, expanding the sources of public funds available to invest in infrastructure, including broadening the use of user fees and applying new tax schemes.
- Third, a willingness by governments to attract and deploy private capital alongside their own infrastructure funds.
- Fourth, measures that can affect greater efficiency in the design and delivery of infrastructure and promote cost-effective innovations, largely based on adopting new technologies.
- Finally, a willingness to expand opportunities to privatize specific infrastructure sectors.

Funding versus financing of infrastructure

In discussing this topic, it is important to distinguish between funding and financing as these terms are not interchangeable. Funding refers to money that is available to use at the time of the expenditure, that is, during the project delivery stage. Typically, this is a government grant or private equity at risk that does not have to be paid back or serviced thereafter. Financing is used when money must be raised or borrowed and ultimately amortized and paid back through some form of debt repayment contract.

Unfortunately, governments around the world have pulled back on infrastructure spending, giving priority to other fiscal expenditures and rising debt fears that even preceded COVID. Many countries, regions, and cities today face the prospect of years of fiscal consolidation and deleveraging to bring public debt down to manageable levels. The only hope is for governments to explore ways to increase revenues including new taxation schemes, user fees, and divesture of obsolete infrastructure. These initiatives will often face formidable political opposition. There was a time that public–private partnerships (PPPs) were seen as an attractive model that could attract private capital to fund new infrastructure projects, but results have shown this not to be the case.

Private investors in infrastructure have three options for investment, only two of which directly address the infrastructure gap:

- Through secondary transactions, invest in existing infrastructure that ties into an operating business, examples being telecom networks, railroads including both the tracks and rolling stock, or deep-water ports.
- Through partnership or co-ownership arrangements, invest in standalone "brownfield" infrastructure projects which may be publicly or privately owned such as airports, transit, windfarms, toll roads, or water and sanitation facilities.

- Venture into "greenfield" projects. To date, most greenfield projects have been initiated by governments, typically using traditional procurement processes. These projects are often not "bankable" from a private investment perspective.

There is a clear preference on the private side—purchase existing projects, either publicly or privately owned, with an established revenue base and with some value-added opportunities. Airports are a good example with the potential growth in retail and cargo activities to supplement passenger travel. The dilemma is that the potential pool of existing projects attractive to investors is shrinking, causing investors to cast a wider net.

Private funding is typically a mix of private equity and commercial debt and involving partnerships or co-ownership arrangements to stabilize long-term operations and underwrite risks. These projects may sit upon publicly owned or controlled land, involve long-term concession agreement, or be subject to some form of regulatory control. The attraction is their long-time horizon, monopoly-like rights that restrict competition, a revenue stream with growth potential, and the opportunity for value creation to drive up returns. Greenfield projects are the least favored by private investors in that they introduce the two risks investors try and avoid—construction and revenue risk.

Leaving China aside, in the U.S., the largest of the investment markets, most private infrastructure financing comes from the corporate sector, usually through corporate bonds and credit lines and not through project finance. Corporate finance makes up about three-quarters of private financing of infrastructure in the U.S. (Edwards, 2017). Maintaining this flow of funds requires regulatory certainty, a sound legal framework to ensure contractual rights, and the ability to charge prices that will produce acceptable risk-adjusted returns. The classic corporate infrastructure investments are in power and communications where they represent 90 percent and 100 percent, respectively, of the investments in these two sectors. Private investors have a much smaller involvement in transportation, water, and sewage—around 10 percent.

Having the political will to close the gap is a necessary, but not sufficient condition, to achieve success. Following through on promises to address infrastructure needs cannot rely solely on private capital and will require an expansion of sources of public funds designated for infrastructure. The three possible options for additional revenue generation in the public sector are user charges, value capture initiatives, and "asset recycling", not all of which may be politically palatable.

User fees have been widely accepted to fund public transit, toll roads, bridges, and tunnels in many jurisdictions. User fees can also be imbedded in the cost of an airline or passenger rail ticket. Value capture is a more complex concept and involves mechanisms by which the public can monetize some of the benefits that will accrue to other than the direct users of new public infrastructure. A well-known application is the £14.8 billion CrossRail project in London, England, to deliver the Elizabeth line east–west railways across the city. An innovative program of finance, funding, and value capture contributed approximately two-thirds of the cost (Buck, 2017). Beneficiaries of this investment were identified as the central business community, adjacent property owners, nearby developers, and Heathrow Airport. All potential beneficiaries were tapped as sources of financial support through a series of separate but coherent measures to raise money and close the funding gap.

Asset or capital recycling—selling existing assets and recycling the proceeds for new infrastructure—has been used in Australia, where the national government established an AU$5 billion incentive program in 2013–2019. The Australian program provided state governments with an additional 15 percent in national funding of the capital raised from recycled assets. Between 2013 and 2016, a total of AU$15 billion was raised in Australia from recycling existing transportation and power-generation assets (US Dept of Transportation, 2020). This is easier to deal with on the real estate side such as the sale of old school sites to reinvestment in new schools at more desirable locations. A famous case of asset recycling in the U.S. is the sale of abandoned military bases, many of which faced significant local and political opposition to base closure.

Australia's experience provides a number of valuable lessons for other governments and private investors, a key takeaway from Australia's experience being that recycling is not always a suitable solution to a country's infrastructure needs (Marshall and McLennan, 2018). Having enough public assets to dispose of to raise sufficient capital is a key prerequisite for an asset recycling scheme. Equally important is the willingness of the public to accept private investment, and in some cases foreign ownership, of infrastructure assets being sold. An example would be the sale of a port or harbor to a foreign interest, a situation that the U.S. government faced with the potential sale of a portfolio of port facilities to a United Arab Emirates-based maritime company (Weisman and Graham, 2006). The sale was blocked by Congress on national security grounds. Asset recycling can be perceived as wholesale privatization of public assets and may cause lasting damage to public perception of this initiative. As appealing as asset recycling may sound, it has not been smooth sailing in Australia, and it has not seen widespread adoption in other countries and even in Australia, post-2019.

Governments have generally not shown a willingness to attract and deploy private capital alongside their own infrastructure funds, except in the context of a public–private partnership, and in relatively small amounts. The PPP model strives to achieve two objectives that are counter to the interests of investors: first, it seeks to minimize the amount of private capital in any project since there is the perception that private money is much more costly than public funds; second, it seeks to minimize the cost of this private money by reducing the spread between government debt and the rate of return on private equity. The result is that most PPPs in developed countries are launched with a very small amount of private capital and at very low rates of return. The situation is very different in the emerging markets where governments are limited in their debt capacity and must rely on development banks, commercial loans, and forms of concessional financing.

Closing the gap is not just about matching money to needs but addressing ways in which needs can be met in more cost-effective ways such as incorporating technological advancement and innovations or introducing new solutions. There are glaring impediments on the construction side where capacity may not exist, or substantial cost overruns and delayed delivery dates have plagued a high proportion of new infrastructure projects. An underlying problem is the lack of productivity in the construction industry, a situation that has prevailed for decades. Productivity growth in the construction sector has been slow or negative in most economies (Vogl and Abdel-Wahab. 2015).

Transferring risks to the private sector, a primary motivation behind PPPs, has not addressed the need for improvements in productivity. In fact, the opposite many be true. Highly prescriptive PPP documents and detailed risk mitigation strategies may stifle innovation and force constructors to continue with business-as-usual practices that they see as far less risky.

New technology can affect both the demand and supply sides of the market for infrastructure and begin to close some parts of the infrastructure gap. Technology also has the potential to address the old adage that the most cost-effective solution is the infrastructure that is no longer needed. One can draw upon examples in other industries: replacement of the neighborhood bank with e-banking; online shopping made possible with new logistical networks and advanced software to facilitate payment and fulfilment; working from home with the support of high-speed Internet services; telehealth; forms of education that no longer require a schoolhouse; and streaming services that have reduced the need for movie theaters.

The infrastructure gap will be around a long time if the current trends continue. The gap will widen as needs in the emerging economies increase to meet the pace of urban growth. There is no single solution to the paradox between demand and supply in a market that cannot clear itself. Multiple strategies will be required.

A starting point may be a common understanding of the problem and consensus on the need for new solutions. There must be a seismic shift to utilize the surfeit of private capital chasing infrastructure projects across the globe with revenue-based business models, and with thinking that goes well beyond PPPs. Governments too often see the PPP as a panacea, yet the numbers don't stack up. PPPs can be improved, and this will be important in the emerging economies, but they will remain a very small part of any effort to capture the level of private investments required to meet future needs. On the public side, new sources of revenue must be sought. The alternative is not palatable—essentially depriving future generations of the infrastructure they will need to sustain a quality of life they are entitled to.

Notes

1 Sourced at www.google.com/search?q=minneapolis+i-35+bridge+collapse&sxsrf=
ALiCzsZ4CAyVosBpNaqALDTL7HtkbIEioQ:1672266852750&source=
lnms&tbm=isch&sa=X&ved=2ahUKEwjl-Oi9r538AhX4mnIEHRw8DOkQ_AU
oAnoECAEQBA&biw=1920&bih=1089&dpr=1#imgrc=oHmSXuM7P-_v8M
2 Sourced at https://en.wikipedia.org/wiki/Hurricane_Katrina
3 DBRS Morningstar

5   Challenges in moving forward

In the future, infrastructure will be shaped by very different factors than those that defined infrastructure needs over past centuries. PPPs have given us a glimpse of the future, but they are still a limited-use alternative to procuring public-sector infrastructure in transportation, energy, social infrastructure, public utilities, government offices and facilities, and even specialized services in communications and defense. To a large degree, PPPs have maintained the status quo by replicating traditional solutions.

Infrastructure design, delivery, and financing today are still firmly rooted in silo mentalities and cumbersome organizational structures within governments. These shortcomings are reinforced by the continued use of outdated technologies whose roots can be traced back to the late 19th century, and by ideological mandates that draw sharp distinctions between the respective roles of the public and private sectors. Infrastructure today is struggling to keep up with growing demands. It is fraught with issues of accelerating functional and technical obsolescence and suffers from the widening gap in the use of emerging technologies between the public and private sectors. Governments are also hampered by institutional resistance such as traditional recruitment practices to secure talent or management hierarchies predicated on seniority. Some of these same governments defend the position that much of our infrastructure is a social good and therefore cannot be trusted to the private sector, nor be subject to user fees. All of this adds up to more constraints on the ability to address future infrastructure needs with creativity, innovation, and new business models.

What is different about future needs when compared to past solutions that requires us to build a different case for infrastructure going forward? The top five most pressing issues include:

- The reality that most of the world's population are now urban dwellers, and the pace of urbanization continues to accelerate across the globe. This is in sharp contrast to rural life that existed for most of the world's population right up to the 21st century.
- The havoc that this urbanization is wreaking upon the planet and the environment upon which we depend. Our carbon footprint and propensity to

DOI: 10.1201/9781003396949-7

foul the air we breathe and the water we drink are contributing to a state of environmental degradation that is not sustainable, even in the short term.

- The inequality that exists across the globe in consuming rapidly depleting resources, geared toward sustaining a lifestyle for those who live in developed societies, and to which all others in emerging economies aspire to achieve.
- The realization that a significant portion of the world's current infrastructure, some of which may date back more than a century, is crumbling, and beyond repair.
- Lastly, the realization that we are on the threshold of a technology revolution that will transform businesses in the decades ahead and may make significant parts of our existing infrastructure networks functionally obsolete, as well as challenge some of the monopolistic powers that governments have carefully crafted and protected.

Responding to these drivers requires more and better infrastructure, and yet most governments are spending less of their GDP on infrastructure and contributing to an ever-widening infrastructure gap. Some would argue that governments have limited options other than to turn to the private sector to provide consumers with choices for the timely delivery of integrated services.

Growth of cities

It is difficult to have a substantive discussion on infrastructure without addressing the growth of cities and the far-reaching sustainability challenges caused by the increasing pace of urbanization across the globe. We have become a planet of cities and live in an urban world where consumers in large cities will account for 55 percent of the global population by 2030 and generate 81 percent of global consumption and 91 percent of global consumption growth (Dobbs et al., 2016). Today, some 55 percent of the world's population—4.2 billion inhabitants—live in cities. This trend is expected to continue. By 2050, with the urban population more than doubling its current size, nearly 7 out of 10 people in the world will live in cities (World Bank, 2020).

Cities are responsible for 75 percent of the world's energy use and produce more than 80 percent of all greenhouse gas emissions (World Bank, 2020). This fundamental change in where and how we live is driving a "shift to sustainability" that incorporates competitiveness along with finding ways to reduce impacts of urbanization on the environment and seeking significant improvements in quality of life for all urban dwellers. This change was described in a communiqué issued by the G20 countries in April 2009. Among the various commitments outlined in the document, was the agreement, "to make the best possible use of investment funded by fiscal stimulus programmes towards the goal of building a resilient, sustainable, and green recovery. We will make the transition towards clean, innovative, resource efficient, low carbon technologies and infrastructure" (London Summit, 2009).

The infrastructure decisions that we make today could radically alter urban growth patterns and could positively influence up to 70 percent of our ecological footprint (UN-Habitat, 2008). These decisions are not just about adapting new technologies, but must be broadened to encompass factors of administration, legal and personal security, and social cohesion (UN-Habitat, 2008). Infrastructure has lagged the technology revolution now underway in most parts of the world (KPMG, 2016). Except for the telecom and energy sectors, there have been few fundamental changes in the type of infrastructure we build compared to that 50 years ago.

A common carryover of our infrastructure today is the lack of stakeholder involvement and end-user engagement to enhance the consumer experience. The missed opportunities to secure user support cannot be underestimated. Nor can the sustainability challenge be met without securing a Social License to Operate (SLO) for new infrastructure projects[1]. A glaring example of this requirement was the failure of Google's Sidewalk Labs to secure support for its proposed Smart City initiative on reclaimed lands along Toronto's waterfront (Jacobs, 2022). In its penchant to pursue new forms of urban infrastructure and institute a whole new level of data gathering and monitoring, Sidewalk Labs forgot about the very people it was intending to serve. This was a fatal flaw in the thinking of what were reputed to be some of the world's leading urbanists and technology experts.

However, the technology revolution is accelerating and redefining the role of infrastructure in shaping our cities. In the renewable energy sector, solar and wind power technology are not only redefining the way that we generate and distribute electricity, but are also disrupting the centralized generation and distribution models that underpin vast investments in the developed world's energy investment strategies. Water distribution and sanitary and solid waste collection based on centuries-old technology is still used in developed economies and is wanting of innovation in developing countries where one can expect to see "leapfrogging" through the utilization of new technologies. Autonomous cars, and trucks, are gaining a foothold in the developed world, however they are met with legitimate public safety concerns and issues of privacy and overreach when dealing with big tech firms. Some of these innovations are beginning to challenge traditional transportation assumptions, as in the case of individual vehicle ownership vs carshare/rideshare platforms and expansion of the "sharing economy" into other related sectors.

Incremental change

Perhaps the most promising role for infrastructure in addressing the sustainability challenge will be relatively small-scale interventions driven by individual consumer decisions with the desire for more direct control over what, when, and how they consume everything from power to energy and transportation. This extends to investing in more smart home technology, discreet personal health monitoring devices connected to telehealth services for an aging population, universally accessible on-line synchronous education, and increased connectivity to enable more robust work from home or "work anywhere" options.

It will be at the nexus between broad infrastructure interventions and the "micro" decisions driven by consumers where new technologies will arise, new business models will evolve, and society will learn to do much more with far less (KPMG, 2016). To cope with the pace of city growth in the decades ahead it is not about technology per se, but about the ability to make informed long-term decisions about the infrastructure we need. This infrastructure must harness the potential of new technologies, both at the micro and macro levels, and derive new business models to capture the full value of this technology. It must firmly place the consumer in the driver's seat.

Infrastructure is seldom a static asset, intended for a specified time. It is better defined as part of a dynamic process reflecting technological advancements, as well as economic, social, and political shifts in society. What defined an interstate highway system back in the early 1900s bears little resemblance to how this interstate network functions today as a primary trucking network and in the future when autonomous trucks may dominate our highways. Social infrastructure will continue to evolve with the needs of a changing demography. The healthcare system globally has been overburdened by the COVID-19 pandemic and the consequences have yet to be fully realized. What are the immediate challenges facing our healthcare systems? How will the concept of care and universal access to this care be affected?

Health care is a critical public service that relies on a complex relationship between long-lasting fixed capital investments in hospitals and care centers, changing medical technologies, pharmaceutical advancements, and added services to drive improvements in preventative care, efficient healthcare delivery, and improvement of the patient experience. Healthcare delivery is already being transformed in a digital era where remote delivery of health care, the use of robotics, and personal diagnostic equipment are rapidly becoming a reality. An unstable public policy context underscores the level of uncertainty and the potential disruption that prevails throughout the healthcare system (Barlow and Köberle-Glaiser, 2009).

During the 1960s and 1970s in the UK, several innovative design and construction solutions were introduced to promote adaptability and "future proofing" of hospitals. Subsequent refinements focused on economies of scale, predictability of costs, and quality control through standardization of construction systems. But with the advent of the Private Financing Initiative (PFI) in the late 1990s the momentum for innovation was thwarted in the interest of standardization. However, by the 2000s adaptability and innovation were back on the agenda, and the Treasury Task Force on PFIs stipulated that new PFI schemes should encourage built-in flexibility and "modern production processes" to enable response to changing requirements with minimal disruption (Barlow and Köberle-Glaiser, 2009).

Results did not bear out progress and some of the blame rests with the PFI model itself. The PFI bidding process stifled innovation when the emphasis shifted to controlling costs and meeting deadlines, and the client was relegated to the role of tenant in their own building, beholden to a consortium of public entities, private developers, and financiers (Barlow and Köberle-Glaiser, 2009).

We may see something similar occur with schools, an anchor for the traditional concept of neighborhood. In many cities, public schools built during the 1960s and 1970s sit vacant as household size became progressively smaller. Families had far fewer children, migrated from older neighborhoods to the outer fringes, and options for private education became more attractive. Recently, the digital era has ushered in the advent of on-line modes of instruction. The Internet combined with personal computing devices that are prevalent in most homes are already challenging the role of dedicated school buildings and sprawling university campuses. How much should we be spending on bricks and mortar versus software, programming, access to high-speed Internet, and computing devices for every household? These changes may also influence our attitude to the use of public buildings and spaces and other forms of social infrastructure that are often underutilized outside limited hours of operation.

Intangible infrastructure

We can see a paradigm shift in investing in infrastructure in Seoul, Korea, where local government has invested heavily in a high-performance digital infrastructure, the backbone of which is the world's highest full-fiber penetration rate (McLaren and Agyeman, 2015). Table 5.1 compares Korea to several other developed countries and the comparisons reveal the country's startling advantage. The investment in this digital infrastructure supports a significant commitment to the sharing economy and has built public confidence and trust in pioneering better use of the social infrastructure in Seoul. For example, through access to high-speed web-based sharing services available to every household, local government is facilitating the use of underutilized public buildings as well as dedicated space in new residential buildings for an assortment of activities including lending libraries, day care centers, tool rental and repair centers, woodworking programs, incubator space for start-up companies, and public meeting and event space. In Seoul, this new infrastructure

Table 5.1 Fiber Penetration Rates

FULL-FIBER PENETRATION RATES (%)	2019 Q2	2015 Q2
Korea	81.65	68.7
OECD	26.84	17.9
United States	15.55	9.7
Canada	15.09	5.3
United Kingdom	2.33	0.3

Source: OECD broadband statistics update, March 2020.

is pushing the boundaries of urban innovation and building partnerships to support flourishing social enterprises.

Korea's impressive and forward-thinking penetration of fiber networks illustrates another aspect of the infrastructure story, that of intangible assets. Infrastructure that refers to things that are intangible encompasses a broad spectrum which includes rules, norms, common knowledge, and institutions (Haskel and Westlake, 2018). This fits within the definition of infrastructure which is costly to produce, durable, and tends to have a public and social character, and makes the economy more productive. The extent to which the growth of intangible investments may radically shift the kind of infrastructure the economy needs is unclear. Proponents of the knowledge economy debate those on the other side who can claim that investments in energy, transportation, and logistics will always be the dominant asset.

The insightful in this debate will want to explore the connections between investment in telecom infrastructure and connectivity and productivity, realizing that new technological infrastructure will be most useful in conjunction with new ways of working. In the 19th century, with the introduction of electrical power, the American factory had to radically restructure its operations to gain the full benefit of this new power source (Haskel and Westlake, 2018). It was a slow transition with just slightly more than 50 percent of mechanical drive capacity electrified some 40 years after the development of the first central electrical plant (David, 1990).

A simple example of transition to intangible infrastructure is the cell phone apps now widely used to manage parking lots and structures. These apps replace lift barriers susceptible to damage, control consoles at entry and exit points, and ticket payment booths and machines. They are convenient, user-friendly, offer more payment options, and eliminate costly maintenance and repair of equipment.

Throughout history infrastructure has meant building what we can see and touch, but investment in "things" may matter less and less to modern economies (Haskel and Westlake, 2018). Intangible investments can include design, research, software, and branding. These investments are scalable and promote more rapid growth of organizations of greater size than tangible investments. Intangible investments can also utilize excess capacity in the system, draw upon assets not normally employed, and build upon existing small businesses. Telecom infrastructure will matter more in an intangible economy to build connectivity between people, and between businesses. It may also herald new solutions to address some elements of the infrastructure gap in the years ahead, as well as disrupt long-term solutions now in place.

We will need to advance the conceptual framework that incorporates and reconsiders all our infrastructure assets to close the widening infrastructure gap, and address a different world dominated by large cities and the quickening migration of people into these cities. Governments will need assistance from the private sector to address sustainability imperatives in response to climate change and provide accommodation for more people within existing urban footprints. They will need the expertise of the private sector to harness the full potential of emerging technologies. Cities can be reshaped to better serve society by expanding the definition of infrastructure and creating new opportunities to engage the public and private sectors in different business arrangements. This can attract pools of private capital needed to invest in the future.

Infrastructure always has been associated in varying degrees with serving the public interest. The private sector must recognize the public interest in approaching infrastructure acquisitions in a framework of business thinking. Governments must move beyond their proclivity to address their long-term infrastructure in a political context. A middle ground is required to bring the two into alignment. Breaking down the ideological barriers that maintain the separation between public and private entities will not be easy to overcome, despite the ever-expanding infrastructure gap that plagues even mature economies. A starting point is recognizing that any public–private arrangement must be founded on business principles that can assure the private sector that it has sufficient resources to endure over the long haul and that it can make a return on its investment. All stakeholders need to be satisfied, public as well as private.

Note

1 Social License to Operate simply means gaining support for a project among concerned groups or various stakeholders over and above what is legally required. This "license" can convey legitimacy or social and environmental responsibility.

Part II
Investing in infrastructure

6   The appeal of real assets

The history of public transportation in U.S. cities, dating back to the 19th century, is an interesting portrayal of the contest between private and public interests in providing an essential public service. Throughout the late 19th and early 20th centuries the growth of urban street railways was closely tied to real estate development and land speculation. Each line extension brought new land within commuting distance of the employment core, sharply raising real estate values. Transit was largely a real estate play controlled by private interests. By the late 1890s, mass transit had become indispensable to the life of large American cities, structured as private enterprises and designed to maximize return for shareholders. One of the well-known examples is the "Main Line" connecting a series of suburban centers to downtown Philadelphia[1]. This line operates today as part of regional public transit system (Schrag, 2002).

Unlike horse-drawn trolleys, both cable-car and electric-streetcar systems required substantial capital for the power plants, maintenance shops, tracks, electrical conduits, and rolling stock. Seeking economies of scale, entrepreneurs formed syndicates to buy up horsecar companies and their franchises, and, when necessary, bribed local governments (Schrag, 2002). "Traction magnates", in Philadelphia, New York, and Boston transformed the industry from one based on monopolies on individual routes, to one based on near or complete monopolies covering whole cities. Many behaved as true monopolists, packing their street and subway cars with riders who had no other choice of transportation. In taking over small companies, the barons also took on enormous corporate debts and sold watered-down stocks to the unwary, leaving the new companies with shaky capital structures.

Many of these private lines were subsequently purchased in the early 20th century by a new set of private players associated with the burgeoning automobile industry. Once purchased by private companies, many transit routes were promptly closed to promote increased reliance on the automobile[2]. In 1942, American automobile manufacturers suspended the production of private automobiles in favor of war materiel. Left without an alternative, Americans turned to mass transit in record numbers. World War II was the last hurrah

DOI: 10.1201/9781003396949-9

for privately operated transit in the United States. Transit ridership peaked in 1946.

Following the War, ridership quickly collapsed. Not only were private cars available and affordable, but so were suburban houses, built so far from central employment areas and so sparsely scattered that mass transit was simply not feasible. Moreover, the construction of new roads, including federally financed expressways, encouraged automobile commuting, whether by driving alone or in a carpool. As a result, transit ridership was no longer profitable and defaulted back to public ownership. Ridership dropped from 17.2 billion passengers in 1950 to 11.5 billion in 1955. By 1960, only 8.2 percent of American workers took a bus or streetcar to work (Schrag, 2002).

The demise of public transportation in the U.S. illustrates two features of infrastructure that highlight the competition between public versus private interests that influence the appeal of these assets even today: the impact of externalities and the role of monopolies (Glaeser and Poterba, 2021).

Externalities

Politicians have historically turned to increased public spending on infrastructure to promote job growth, counteract the consequences of a recession, or address productivity concerns. An argument for increased spending on public infrastructure is infrastructure-led economic development, emphasizing the success of the Tennessee Valley Authority (TVA) following the Great Recession in the late 1920s. The TVA raised average incomes by shifting employment from agriculture to manufacturing, driven by access to cheap electricity. Whether new infrastructure can substantially increase economic activity in poorly performing regional economies is still debatable, despite what politicians may say. It is difficult to assess the long-term macro-economic effects of infrastructure projects given the potential range of confounding factors (Glaeser and Poterba, 2021).

Infrastructure has externalities that offer a sound rationale for government intervention. Local governments dating back to the 1800s invested in water infrastructure to prevent the spread of disease in large cities, often under the auspices of an independent public authority. Public sewage had an even higher ratio of public benefits and the combination of clean water and public sewerage pushed down death rates. While local externalities are effective arguments for the provision of clean water and public sewerage, the arguments in favor of public ownership of clean water and sewerage are no longer as convincing as they once were. Public water systems have fallen into disrepair, subject to massive leakage problems, and not always reliable when it comes to meeting quality standards.

The famous case in the U.S. is the Flint, Michigan, water crisis, a human-made, modern-day public health crisis (April 2014–June 2016) involving a municipally owned water supply system. As a result of some cost-saving measures by local government, tens of thousands of Flint residents were exposed to dangerous levels of lead, and outbreaks of Legionnaire disease killing at least 12 people and sickening dozens more[3].

The argument can be made that private water systems can offer clean water at competitive rates and be subject to regulation and constant public monitoring. Similar arguments can be made for private sewerage. Water privatization in France, often as public–private partnerships, goes back to the mid-19th century when cities signed concessions with private water companies for the supply of drinking water[4]. As of 2010, according to the French Ministry of Environment, 75 percent of water and 50 percent of sanitation services in France were provided by the private sector, primarily by two firms, *Veolia Water* and *Suez Environnement*. However, private ownership of these fundamental assets it is not without its critics. A comparative assessment of public and private service provision is complicated by the absence of a mandatory national performance benchmarking system. Supporters and opponents of private sector participation often find it hard to provide objective figures to back up their respective positions.

There is a tipping point on privatization, and it may have recently been reached in England where the Guardian newspaper reports that 70 percent of the English water industry is in foreign ownership (Leach et al., 2022). Most of the water sector is owned by a complex web of investment firms, wealth funds, and tax haven-based businesses. Ownership structures are so complex that transparency and accountability are opaque. The U.S., for example, has the strongest foothold in English water companies, with investment firms owning nearly 17 percent overall. Canadian and Australian companies are the second and third biggest overall investors in English water. At least a fifth of the industry is owned by corporations based in Asia. The tipping point is reached when the public interest is not being served due to lack of investment to maintain this infrastructure and public health and safety issues arise. This highlights the issue of offshore ownership and the motivation to drive up global returns at the expense of local health and safety concerns.

Transportation systems offer complex arguments on public versus private involvement. For example, when it comes to assessing externalities associated with transportation systems, surface congestion is often the cited example. Public transit requires a user fee in addition to public subsidies, whereas most roads, other than toll roads, are fully subsidized. The result is that public transit in many cities has declining ridership and increasing deficits, while traffic congestion intensifies. The politically expedient response to traffic congestion is ramping up public spending on more roads and road widenings. The externalities of this intervention include rising air pollution levels, lost time, loss of valuable land, and disruption of existing neighborhoods. Seldom will the public sector explore alternatives to reduce these externalities such as increasing subsidies to public transit, applying congestion charges, or introducing road tolls. Toll roads are one of the few areas where governments have turned to the private sector. Government will argue that roads have social and economic benefits, while critics will focus on the negative externalities of facilitating greater use of the car.

Monopolies

Some of the distinctive qualities of infrastructure make it difficult to maintain competition among infrastructure providers, leading to what is referred to as a *natural monopoly* (Gómez-Ibáñez and Liu, 2022). Monopolies are common in the public provision of infrastructure and serve as the basis for many concession agreements, whereby a government transfers some or all its monopoly to a private provider. Monopoly control is reinforced by economies of scale that allow one provider to dominate an entire market. A regional toll road or a connecting bridge over a body of water are examples. Adding to this are the durability and immobility of infrastructure investments that are an effective obstacle to competition, reinforced by the absence of any cost-effective substitutes.

How best to protect consumers with remedies that can address the lack of competition is a matter of intense debate (Gómez-Ibáñez and Liu, 2022). This is of great public concern with the rise of consumer reliance on digital networks that support everything from the Internet to cell phones. Public response to natural monopolies, or oligopolies, can take one of two forms: regulation of private providers, or the introduction of a public operator, for example, a state-owned enterprise (Glaeser and Poterba, 2021). Unfortunately, there are limited options to address the competition problem in certain infrastructure sectors and publicly condoned monopolies and oligopolies are prevalent in many countries.

Rationalizing the public interest in infrastructure assets

It is easier to defend a rationale that governments should be responsible for providing essential infrastructure than it is to find reason for private intervention into some of this infrastructure. Glaeser and Poterba cite three reasons for drawing the public sector into the ownership and operation of infrastructure assets: (1) when the scale of investment is thought to be too large for private investors; (2) when positive externalities that would not be considered by private investors exist; and (3) addressing concerns that infrastructure could be used by a monopoly to exploit those who require this service.

While the historical rationale for public investment in infrastructure has diminished over time, many of the reasons for public intervention remain. What we have seen in recent decades is an increasing presence of the private sector in infrastructure through privatization of existing assets, the provision of new infrastructure under private ownership, the transfer of public infrastructure from the public to the private sector through concession agreements, and other contractual arrangements including outsourcing and public–private partnerships. The increasing private sector presence in infrastructure can be attributed to the distinctive nature of these investments, coupled with the rise of infrastructure capital that is a source of funding for large-scale infrastructure projects. This increasing presence is not without its detractors and raises significant issues when it comes to assets that have a distinct public purpose such as foreign ownership of water systems in England, as mentioned.

Evolution of the capital markets

Capital markets are no longer underdeveloped in the Western world as they were in the early 19th century when governments carried a significant financial burden for new infrastructure. In many countries, including the U.S., the public sector was once the only source of funding for large-scale infrastructure projects. In comparison, by the end of the 19th century, capital markets in the UK were well enough developed to support private investments in infrastructure. It helped that projects in the UK were not as capital intensive as in the U.S. as infrastructure involved flatter surfaces and shorter distances. In early times, despite being public funded, the general assumption was that users should still pay for this new infrastructure.

The Erie Canal, previously discussed, linked the Hudson River to the Great Lakes and was publicly funded but user charged. The Canal was so successful that user fees quickly recovered costs (Glaeser and Poterba, 2021). Benefits extended well beyond the actual users of the Canal as evidenced by the prosperity enjoyed by cities such as Buffalo, Syracuse, and Rochester in upstate New York. This example harkens back to Adam Smith and the *Wealth of Nations*, published almost 50 years before the Erie Canal was built. Adam Smith extolled the virtues of user-funded infrastructure projects: "When high roads, bridges, canals, are in this manner made and supported by the commerce which is carried on by means of them, they can be made only where that commerce requires them, and consequently where it is proper to make them"[5]. Today we find a situation where the capital markets are searching for good investments, the public coffers are all but depleted, and there is a hesitancy among many governments to fund public infrastructure with user fees.

Desirable features of infrastructure as an asset class

For investors, infrastructure and real estate are closely related asset classes that have similarities, but also have their own special features that set them apart. Infrastructure has unique characteristics that enhance its appeal to long-term investors, and particularly large institutional investors. Gomez and Liu identify five characteristics, the combination of which make infrastructure attractive to investors (Gómez-Ibáñez and Liu, 2022). The five characteristics are: essentiality; costly, durable, and immobile; difficult to store; equity; and externalities.

- **Essentiality**. Infrastructure is essential to the delivery of almost all goods and services in the economy and society. Infrastructure affects quality of life; the economic development and competitiveness of cities, regions, and countries; reduces the cost of consumption and production; and opens up opportunities for global trade and commerce. Infrastructure is always high on public policy agendas. Investment in infrastructure, as a percentage of GDP growth, is seen as a proxy for the welfare of citizens and the economic development of nations.

- **Costly, durable, and immobile.** Infrastructure is expensive to build and operate, lasts for a very long time, and cannot be moved once installed. There are a few exceptions. These three factors also introduce some significant risks for investors in terms of the impacts of depreciation, functional obsolescence, and the possibility of owning a "stranded" asset that no longer has any economic value. One is reminded of the extensive infrastructure that supported the ice industry in 19th-century New England including ice storage sheds, transportation equipment, ocean-going ships, and ports that served international destinations[6]. The ice industry once represented a significant component of U.S. GDP in the 19th century, employing more than 90,000 people. Within a very short period, due to the introduction of the compressor and refrigeration, the entire infrastructure supporting the cutting, storage, and global shipping of ice was rendered obsolete (Hahn and Laskowski, 2002). Durability can also be an impediment to introducing competition and technological advancements, plus increase the likelihood of deferred maintenance.
- **Difficult to store.** Most infrastructure is not known for its storage capabilities, other than water storage in reservoirs. Storage systems are expensive to build and maintain. The need to store electricity is a significant impediment in the energy business and impacts a range of industries. Lack of storage capacity leads to challenges in meeting demands in both space and time and particularly peak hour estimates, inefficiencies in pricing, and consistent quality in service delivery. There are few solutions to redress the storage problem other than recent advances in technologies such as battery or hydrogen storage in the energy sector.
- **Equity.** This is a significant issue for infrastructure and almost all infrastructure involves equity consideration. Does a project provide for services that are widely available, affordable, and inclusionary? Does a site or a project burden minorities, indigenous people, or other disadvantaged groups? What standards of availability should apply? Does a proposed project have community support or strong opposition? Should certain users be subsidized? Are there social and environmental impacts to be considered? Each project will have its own equity issues with the expectation that ultimate resolution is a public responsibility. The equity issue underscores the importance of securing and maintaining a "Social License to Operate" for large infrastructure projects.
- **Externalities.** As previously mentioned, a distinct characteristic that sets infrastructure apart is the externalities involved. These range on the positive

side from job growth to improved services, and on the negative side from global warming to destabilizing a wetland. Infrastructure can enhance or destroy land values, imperil or strengthen neighborhoods and entire communities, reinforce patterns of agglomeration, or shift directions of growth such as when an airport gets built. These externalities raise questions of assigning some of the costs to beneficiaries, implementing value capture by selling development rights, or imposing local improvement taxes. These measures often spawn political debates as there are inevitably winners and losers in any infrastructure project.

Challenges and risks in investing in infrastructure

Infrastructure risks fall into two categories: those that are specific to infrastructure as an asset class, and those that are specific to a particular asset. Risks common to the asset class are primarily political and regulatory, while asset specific risks include construction risk, operational risk and financing risk (Deutsche Asset Management, 2017).

Political and regulatory risk: This also includes sovereign risk, particularly in emerging economies where the threat of nationalization or political instability may exist. Given the importance of infrastructure to a well-functioning economy and to the well-being of citizens, governments will

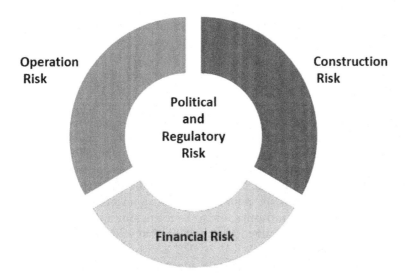

Figure 6.1 Risks common to the asset class.

Source: Deutsche Asset Management, March 2017.

find ways to maintain some level of control or regulation over infrastructure assets to protect the public interest. Political and regulatory risk will vary across countries and regions and across sectors, and can have significant impact on long-term investment returns.

Construction risk: This risk is a primary reason why investors avoid green-field projects and prefer mature assets. It is not a risk that investors can easily manage or mitigate. Construction risk includes cost overruns, delivery delays, design errors, or changes in scope or specifications during construction. Construction risks can also embody local labor practices, available skilled labor, and ready access to materials and assemblies at competitive costs. These risks are particularly high for large, complex projects and are compounded by the introduction of new materials, new technologies, and low productivity in the workforce. The consequences can include postponement of a project, higher capital costs, delayed revenues, and reduced returns.

Operational risk: When fully operational, infrastructure assets are subject to performance risk and maintenance risk. Performance shortfalls can include failing to meet volume or revenue targets, and unanticipated cost increases. Maintenance risk can include unplanned maintenance needs, systems failures, and unexpected shutdowns. Exposure to maintenance risks will vary by asset type. Some regulated assets allow cost pass throughs to customers to address operational setbacks, whereas unregulated assets imply increased cost for the owner.

Financial risk: This risk will depend upon the capital structure of the asset, including the level of leverage and debt. Some financial risks can be mitigated through hedging, but usually at a high price. Financial risks include increasing interest costs, currency fluctuations, exchange rate risk, and a higher cost of debt for cost overruns and other capital calls.

The challenges facing investors in infrastructure are three-fold (Tian et al., 2020):

Limited liquidity: Infrastructure involves long-lived immobile assets that are considered illiquid because of their limited trading options in secondary markets or binding contractual obligations, especially with PPPs. Those investing in infrastructure require knowledge in operational models and management structures, plus an understanding of the risk exposure of a particular asset. This limits the pool of prospective investors and directs acquisitions on large-scale infrastructure to large institutions who typically deal in direct investment, or limited pooled investment vehicles that consolidate funds from smaller investors.

Transaction inefficiency: Infrastructure investments often have high friction (transaction) costs associated with bespoke direct investments, fund management overhead, and other fees and expenses. There can also be costs incurred by the involvement of multiple intermediaries. Many infrastructure assets are unique and involve complicated legal agreement to ensure operating efficiency, to clarify the roles and responsibilities of the various parties involved, ensure the proper flow of revenues proceeds, and confirm the various risk-sharing mechanisms. These costs can be minimized by large institutional investors pursuing direct investment and capitalizing their transaction over long holding periods.

Lack of transparency: The infrastructure universe is characterized by information asymmetry starting from the procurement process through to operations and maintenance. Information sources can be unique to the asset class, highly scattered and inconsistent, and not easy to access. This lack of transparency introduces the possibility of corruption, inadequate governance, and unfair competition in some markets. This can dampen the enthusiasm of investors interested in this asset class.

Impact on pricing

All of the above will impact decisions by governments and investors on whether to proceed with a project, and at what price (value). Pricing will influence the decision on whether an infrastructure project should be private or public, or some combination of both, what capital cost is justified, the operating costs, and who should pay to cover some or all of these costs. From a public perspective there must be clear public benefits, risk mitigation at reasonable costs, and some assurances that the right choices are being made starting with a competitive procurement process. Investors will want to know whether there is sufficient revenue to motivate private investment and certainty of net cash flows to determine asset valuation. Cashflow modeling is an essential part of any pricing exercise. Pricing influences investment decisions, determines level of usages, accounts for depreciation, and may include incentives for improved performance and levels of customer satisfaction. Pricing will be a defining factor in determining the appeal of a given infrastructure investment to an investor.

The appeal of investing in infrastructure

This brings us back to a fundamental question: Why do private investors invest in infrastructure? This is an asset class with its own unique challenges, significant risk factors, and subject to debate on who should own what infrastructure, and why. Even defending what the term infrastructure includes can be an exhausting exercise. Investors have different expectations for infrastructure as compared to their expectations for private equity, key reasons being a desire for value creation, reliable income, and inflation hedging.

In comparison to private equity transactions, with expectations of high absolute returns and high risk-adjusted returns, infrastructure offers more diversification benefits, reduced volatility, and a better match for their long-term liabilities. These are the characteristics that the asset management industry focuses on. Regulators have also recognized the benefits of the asset class. In 2007, amendments to the EU Solvency II Delegated Regulation on insurance firms reduced the amount of capital which insurers must hold against equity qualifying as infrastructure, and in 2020 the investment regulation of Swiss pension funds (BVV2) separated infrastructure from other asset classes, allowing an allocation of up to 10 percent to infrastructure a joint cap maximum cap of 15 percent on all other alternative investments (Andonov et al., 2021).

Notes

1 Sourced at https://en.wikipedia.org/wiki/Philadelphia_Main_Line
2 General Motors, Standard Oil, and Goodyear Tire formed a consortium during the 1930s whose sole purpose was to purchase urban light rail systems and scrap them. This forced cities to purchase buses from GM, fuel from Standard Oil, and tires from Goodyear. After WWII, they used their clout with Congress to keep stifling regulatory requirements on the rail transportation industry in place, while at the same time pushing for increased expansion of the U.S. highway system. Sourced at www.quora.com/Is-the-auto-industry-the-reason-why-most-cities-in-the-US-don-t-have-proper-public-transportation
3 Sourced at www.britannica.com/event/Flint-water-crisis
4 Sourced at https://en.wikipedia.org/wiki/Water_privatization_in_France
5 Sourced at www.cato.org/blog/adam-smith-infrastructure
6 In 1853, Alexander Twining was awarded U.S. Patent 10221 for an icemaker. In 1854, James Harrison successfully built a refrigeration machine capable of producing 3,000 kilograms of ice per day. In 1867, Andrew Muhl built an ice-making machine in San Antonio, Texas, to help service the expanding beef industry before moving it to Waco in 1871. Sourced at https://en.wikipedia.org/wiki/Icemaker

7 Investing in infrastructure

A wall of money

The first big wave of investments in infrastructure dates back to the first wave of privatization in the 1980s when large firms associated with the industry such as GE, Siemens, and ABB launched their own private equity funds to take on the early arrival of concession agreements, raising both equity and debt (Blecher et al., 2019). This was followed by the bank-sponsored funds in the 1990s and 2000s, raising monies from institutional investors to deploy into core infrastructure assets. The third wave represents a very different approach to investing in infrastructure. The focus shifts away from simply deploying capital, to adding value-transforming infrastructure companies by investing in growth, improving efficiencies, and driving up returns with risk mitigation strategies (Blecher et al., 2019). The result was enhanced risk-adjusted returns that attracted an increasing number of large institutional investors. Investors today see themselves as owners of infrastructure companies. They have a private equity mind set, are operationally focused, and know the importance of having qualified people in the right positions with the relevant experience.

The modern era of private-sector involvement in public infrastructure began with the wave of privatizations in the 1980s and 1990s that fostered the PPP/PFI boom in the late 1990s and early 2000s in Australia, the UK, then the EU and Canada (Mitchell and Job, 2019). The growth of PPPs was paralleled by an increasing interest in real assets by institutional investors looking for suitable long-term assets. Distressed commercial real estate assets in the early 1990s offered such an opportunity. These assets, many of which were in mature markets and under the control of commercial lenders, were readily available for acquisition at discount prices. Prior to these acquisitions, institutional investors lacked familiarity with real assets and were not comfortable taking on operational and market risks. They remained cautious and allocated only a small fraction of their investments to real estate and almost nothing to infrastructure.

The need for infrastructure continued to grow and efforts were launched to develop new financial instruments and techniques for infrastructure financing.

DOI: 10.1201/9781003396949-10

Early responders relied on expertise in structured finance such as used in the leasing industry where equipment leases were securitized using the principles of project finance, as well as in providing infrastructure for the mining, oil, and gas industries. These were assets with agreements in place that could support the long-term financing of capital-intensive large-scale projects. Developments in the equity market for infrastructure held promise and the creation of a liquid market for project bonds proved to be a desirable complement to syndicated loans to finance projects.

This era also gave rise to the notion of direct investment in infrastructure among some of the large Canadian pension funds. Several of these funds, rather than just buying distressed real estate assets and paying someone to manage them, purchased the real estate firms that previously owned the assets and thereby addressed the asset management issue that was always a concern. It was a "two-for-one" deal. Funds secured ownership of assets and at the same time acquired the talent to manage the assets. These assets came at deeply discounted prices and the talent, under a recognized brand name, came at minimal cost. In Canada, three of the big players stepped into this market: OTPP acquired Cadillac-Fairview, OMERS acquired Oxford Properties, and CDPQ acquired Ivanhoé-Cambridge. All three were long-established Canadian real estate firms with national and international operations and a portfolio of high-quality, well-located real estate assets. Experience in real estate built up confidence in real assets and helped launch infrastructure investment platforms in all three pension funds. This initiative vaulted these three pension funds to among the ten largest investors in infrastructure globally, where they remain today.

Growth in the sector

Investors still see infrastructure as one of the three real assets, along with real estate and natural resources. The three are referred to as the "alternative asset" classes, perhaps a misleading term given that these assets are the foundation of most economies. These are assets inherently tied to global trends in GDP growth, inflation and deflation cycles, and various macroeconomic drivers of the economy. They are also assets associated with income streams and cash flows that can be monetized, priced, and transacted in global markets. Real assets represent tangible things that can be seen and touched, and this provides comfort to investors in search of real returns linked to inflation.

Infrastructure assets reflect the desires and preferences of society, the policy and regulatory framework of a country, and the demand for essential goods and services by consumers and businesses. Revenues generated from these assets are considered relatively predictable, sustainable, and capable of withstanding economic cycles. Unlike real estate, which is largely a local business, infrastructure adopted a global perspective right from the start (Henry, 2019).

Over the past two decades, growth in this sector can be attributed to several factors:

- Immense growth potential for infrastructure world-wide. Demand is fueled by both population growth and the growth of cities and intensified by the impacts of climate change, environmental degradation, and the advancement of new technologies.
- Protracted period of low interest rates, when combined with low returns from traditional investments, generated an appetite for higher yielding assets.
- Protection from the lingering hangover from the financial crisis of 2008–2009 and the recognition that little has been done to prevent a repeat of such a collapse (Elliot, 2019).
- The "wall of money" arising from investors across a range of investment vehicles from pension funds and sovereign funds to both listed and unlisted infrastructure funds and the large mega-equity funds[1]. In a single decade, just for unlisted fundraising for infrastructure, the total went from $11.6 billion in 2009 to $100 billion by 2019 (Alves, 2020).

A case for investing in infrastructure

The concept of real or tangible assets contributing to an inflation-sensitive portfolio has been gaining ground in the last decade, and a portfolio mix of real estate, infrastructure, and natural resources has been popular among Australian, Canadian, and EU institutional investors. The large Canadian pension fund, CPP Investments, was among the first to consolidate investment in real assets with the appointment of a Managing Director and Global Head of Real Assets overseeing all investments in commercial real estate, energy and resources, infrastructure, power and renewables, and agriculture land. Similarly, CalPERS, one of the largest pension funds in the U.S. instituted their Real Assets Division in 2016, overseeing real estate, infrastructure, and forestland. This integration has been slower to gain traction elsewhere, but the trend continues to more integration of real estate, infrastructure, and resources under the title "real assets".

For investors, there are significant potential benefits of real asset diversification beyond cash flow stability. Real assets can provide an effective way to enhance portfolio diversification beyond an allocation to stocks and bonds and offer predictable steady streams of income, capital appreciation, higher risk-adjusted returns, as well as inflation protection. In a report issued in 2016, Brookfield Asset Management summarized the potential benefits they saw through real asset diversification (Brookfield, 2016). The five benefits they identified include portfolio diversification, capital appreciation, predicable cash flows, high risk-adjusted returns, and inflation protection.

Portfolio diversification

Because individual real asset categories have been shown to have low correlation with each other, investors can further diversify by investing in more than one real asset class. This explains the involvement of the large institutional investors and private equity funds in both infrastructure and real estate—drivers behind the returns on each are relatively distinct. In Table 7.1, the assumption is that the return streams of two asset classes having a correlation of 1.00 are perfectly correlated. The correlation infrastructure with real estate is 0.55 and low with stocks at 0.64 and bonds at 0.37. The data support the objective of portfolio diversification, with very low correlations between private market infrastructure and the other asset classes shown in the table. Private market infrastructure also has very low correlation with listed infrastructure, suggesting that listed infrastructure is not a good, liquid proxy for private market infrastructure (Lipchitz, 2019). Listed infrastructure offers others benefits.

With infrastructure, the cash flows from some infrastructure assets, such as toll roads, ports, and airports, tend to rise with expanding macroeconomic conditions, while those derived from more essential services, such as utilities, tend to be more highly regulated, and thus more stable in periods of economic downturn. Real estate can exhibit some similarities in that global real estate varies significantly according to the degree of economic sensitivity within a given country.

Capital appreciation potential

Both infrastructure and real estate share in the benefits of potential capital appreciation (Figure 7.1). These sectors represent long-lived, hard assets that tend to increase in value over time as replacement costs rise, cash flows increase, and firms pursue operational efficiencies. This places a premium on assets that are well located in high-barrier-to-entry markets such as a well-positioned toll road with rising volumes, rising demand for energy in regulated markets, and international airports with year-to-year passenger growth.

Table 7.1 Low Correlation Among Real Asset Constituents

	Real estate	Infrastructure	Timberlands	Agriculture	Stacks	Bonds
Real estate	1.00	0.55	0.23	0.12	0.19	(0.10)
Infrastructure	0.55	1.00	0.17	0.08	0.64	0.37
Timberlands	0.23	0.17	1.00	0.64	(0.11)	0.07
Agriculture	0.12	0.08	0.64	1.00	0.09	(0.15)
Stocks	0.19	0.64	(0.11)	0.09	1.00	0.23
Bonds	(0.10)	0.37	0.07	(0.15)	0.23	1.00

Source: Brookfield Asset Management. The Benefit Potential of Real Asset Diversification. 2016.

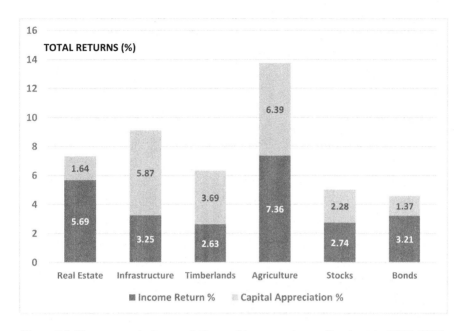

Figure 7.1 Ten-year capital appreciation and income returns of real assets (2006–2016).

Source: Brookfield Asset Management. The Benefit Potential of Real Asset Diversification. 2016.

Predictable and steady streams of income

The cash-flow streams of real assets are often supported by regulated or contracted revenue arrangements and attractive terms to support operating margins. Many of these assets are businesses subject to long-term lease or concession agreements with pricing provisions that ensure predictable returns over time. The Brookfield analysis indicates the 10-year annual growth rate of EBITDA (earnings before interest, taxes, depreciation, and amortization) for infrastructure (2006–2015) was 8.2 percent.

Potentially higher risk-adjusted returns

Real assets—real estate, infrastructure, and real asset debt—delivered relatively attractive long-term risk-adjusted returns, all near or above the capital markets line between the risk-free return (RFR) and equities (EQT) (Figure 7.2).

Inflation protection

Real asset revenue streams often respond favorably to periods of higher inflation, as short-term contractual revenues such as annual rents benefit from

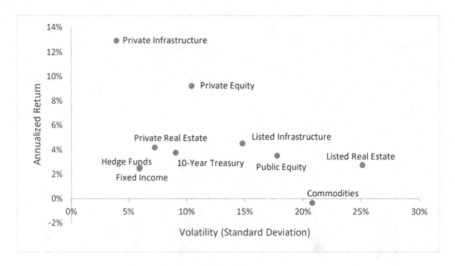

Figure 7.2 Risk and return of asset classes (2008–2017).

Source: Clive Lipschitz and Ingo Walter. Bridging the Gaps: Public Pension Funds and Infrastructure Finance, 2019.

frequent resets and longer-term contractual revenues such as concessions include scheduled rent escalations linked to inflation. There is also the possibility of volume increases during inflationary periods, particularly when driven by strong economic growth and increasing levels of employment and consumption. As a result, real asset returns tend to exhibit greater sensitivities to inflation than traditional investment alternatives (Figure 7.3).

Current status of the asset market

Infrastructure investing is reaching an important juncture as institutional investors allocate more and more capital to infrastructure but find it harder and harder to deploy this capital. There is a clear supply–demand imbalance (Inderst, 2019). On the demand side, unlisted/private infrastructure assets under management constitute about 1 percent of institutional portfolios according to the OECD, a market size that OECD estimates in the range of approximately $500 billion globally (Inderst, 2019). Most investment boards are revamping long-term investment strategies with a higher allocation target to infrastructure. However, the spread is uneven across pension funds. Australia and Canada have allocation targets for infrastructure well over 10 percent and rising. Most of the smaller funds have relatively low commitments to unlisted infrastructure. To highlight the dilemma, if there was a move across all funds to increase their allocation from 1 percent to the 3–5 percent range, this would translate into an increased demand for infrastructure assets in the $2–4 trillion

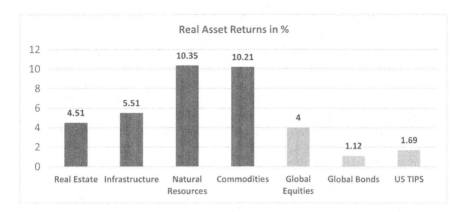

Figure 7.3 Periods of positive inflation surprise (2002–2016).

Source: Brookfield Asset Management. The Benefit Potential of Real Asset Diversification. 2016.

range annually (Inderst, 2019). Supply-side constraints make achieving such increases highly unlikely.

Supply-side impediments include prevailing concerns with privatization, even in many developed countries. In some countries, there is also the potential of nationalization of crucial infrastructure assets such as water, rail, and airports at some future time. Even developed countries may restrict foreign ownership of strategic or vital sectors including seaports, energy distribution networks, and high-tech infrastructure. Cybersecurity concerns are now part of an initiative to reign in the use of foreign technology in national telecommunication networks.

Over the last two decades the market has seen increasing sources of private capital raised through a range of new investment products. Coupled with this is the dominance of large investment funds and the appeal of the Canadian model of direct investing. According to Inderst, for the smaller pension funds, co-investment platforms have been created to overcome lack of scale and overly concentrated exposure.

There is a caveat in discussing the current market and that is the impact of the economic uncertainty that began to take hold in early 2022 and is forecast to prevail for the next few years. Even though the effects of macroeconomic headwinds are now becoming more apparent, investors anticipate strong upcoming performance. The illiquid asset classes—private equity, infrastructure, real estate, and particularly private debt—still represent an increasing proportion of new investment searches by the investment funds.

Need to reassess risk and return

Historically, investors in infrastructure looked at real assets that they can feel and touch, are essential services such as energy or transportation, and can offer steady and stable returns with good downside protection. On the financial side they expected cash yields driven by high EBITDA margins that can provide risk protection and cushion upfront capital expenditures (Brinkman, 2022). Change is happening rapidly. Markets are still in turbulence with the lingering effects of COVID, breakdowns in logistical networks, increasing skepticism over the positive attributes of globalization, and the war in Ukraine. Inflation is at levels not seen for decades, interest rates are rising, equity markets are volatile, and there are threats of a recession in 2023. On the geopolitical front, the ambitions of Russia and China are yet to be fully revealed and the influence of India is on the rise.

Overlay this with the deeper, more gradual, and pervasive way in which infrastructure assets are changing. There are revolutions in energy with carbon-based fuels giving way to wind and solar, plus the rapid ascendency of green hydrogen. Mobility is changing with EVs, battery technology, and "smart" motorways. Digitization is affecting so many things from mass production processes to artificial intelligence, communications, health delivery, education, and entertainment. Cumulatively, these interventions are causing structural shifts in the economy and raising fears of functional and technical obsolescence in some infrastructure sectors. There are also significant economic and social transformations underway rooted in the growth of cities, the aging population, resource scarcity, plus environmental mandates. These change factors add up to disruptions in historic investment patterns, new opportunities for investors, and threats to existing portfolios and risk/return expectations.

In responding to both threats and opportunities, investors can expect to take on more risks, initiate strategies to protect market positions, accept periods of high cash demand and negative cash flows, and acquire new skill sets and knowledge to adapt emerging technologies to commercial uses. All of this translates into the need to reassess traditional risk-based classifications and the risk/return profile of specific asset classes. Infrastructure may no longer be seen as the safe haven it was over the past two decades.

Marcel Brinkman in his McKinsey report claims that investing in infrastructure today and in the future cannot rely on past performance which has been stellar when compared to other asset classes (Brinkman et al., 2022). This implies lowering expectations, and three changes stand out. First is the need to reassess entire asset management practices including a thorough portfolio review with a keen eye to the asset classes that represent new value drivers, trends, and evolving technologies. Second is a consequence of the amount of money being raised, the increasing sizes of funds, and the intense competition for assets to place this money. This is driving up the size of transactions and minimum equity requirements. Third is the recognition that expectations for returns for infrastructure investment are declining and this is a long-term trend.

McKinsey estimates that the average target internal rate of return (IRR), based on low- and high-target IRRs across 20–40 funds per year, and across all primary investment strategies (core, core plus, fund of funds, debt, opportunistic, and value added) has declined in the period 2009–2021 by 4–5 percent

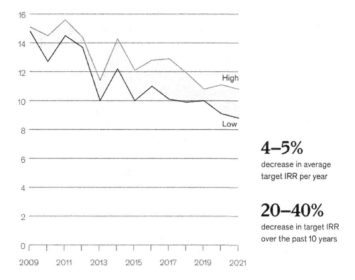

Figure 7.4 Expectations for returns from infrastructure investments, 2022.

per year (Brinkman and Sarma, 2022). This translates into a 20–40 percent decrease in target IRRs over the 10-year period (Figure 7.4).

Impacts of climate change on the investment industry

The impacts of climate change are having a profound effect on infrastructure and shifting the emphasis to the "E" of Environmental, Social and Governance (ESG). High temperatures, never previously recorded, were reached in Britain, Europe, and parts of India and China in July 2022. These temperature levels are a potent reminder that large parts of our built environment, including infrastructure, are built for a bygone era (Economist, 2022). Wildfires are raging, roadbeds are bucking, as are airport runways. People are dying of heatstroke, particularly in countries that have never relied on air-conditioning for residences and places of work. Add to this the damage to agriculture and the severe droughts now threatening water supplies, even in the rich countries. There is mounting evidence of the costs involved and the disruptions in upgrading infrastructure to meet the intensity and frequency of the predicated heatwaves in the decades ahead. The Economist cites the example of welded rails in Britain that are optimized to be stress-free at 27°C, whereas temperatures are rising into the high 30s. Rails can be adapted, but at great cost. Runways can be rebuilt, but not without significant disruptions to air traffic. Air conditioning can be added to buildings, driving up electricity usage.

Countries closer to the equator and subject to intense heatwaves will experience temperatures too hot for human survival. Adaptation in these circumstances is not possible. Parts of India are already coping with the human dimension of climate change as temperatures in several major cities are approaching 50°C, combined with critical levels of air pollution (Watts, 2018). Climate change will be a critical factor in making future infrastructure investment decisions. It will impact current infrastructure that cannot be effectively adapted at reasonable cost, render some infrastructure functionally obsolete, and force investors to impose a sustainability decision-making framework much more rigorous than the current emphasis on ESG.

Note

1 This term can be attributed to Bruna Alves in his Editor's Letter. "Welcome to a decade of infrastructure", Infrastructure Investor. The Decade. September 2019, p. 3.

8   Infrastructure as a business

Governments see infrastructure as a prerequisite to improving delivery of services to taxpayers and users. Investors see infrastructure as a revenue-generating business. Governments focus on the procurement of new infrastructure within a relatively short time frame; investors focus on managing existing infrastructure with a revenue stream over long time periods. These are two very different platforms and help explain the difficulties of bridging the two solitudes.

Governments have a responsibility to provide for the essentials of everyday life, as well as bolster economic activity. They must ensure the uninterrupted delivery of such basic services as electricity, clean water, paved highways and roads, telecommunications, and transportation networks, no matter who owns them. Government services also extend to health care, education, social services, recreation and, perhaps more contentious, the provision of food and shelter for the disadvantaged in society. How they do this is at their discretion. Citizens take these services for granted and expect them to be delivered without interruption, and at the lowest possible price. The need for this infrastructure is also "derived" in that it is seldom desired for its own sake, but required as an input to the delivery of other goods and services. As Glaeser points out, public infrastructure shapes the density and form of cities and has broad public policy implications (Glaeser, 2022).

Governments face public policy decisions as to what infrastructure to build, when to build it, how to procure it, and how to pay for it. Governments focus on decision-making processes that are seldom clear-cut, may be politically motivated, and the tools of investment evaluation are heavily weighted to some form of cost–benefit analysis to include social and economic benefits (Pickrell, 2022). The decision-making process is biased to "*shiny new things*" that garner more positive responses from taxpayers. Investing in maintenance and repair has limited political benefits, and the effects of deferred maintenance are not immediate and not always obvious. Governments are also forced to choose between investing in competing projects as they allocate their limited resources. There are always winners and losers in this public "lottery" process.

Investors prefer to approach infrastructure as owners, usually as partners with other investors or owners in the business enterprises they acquire. These

DOI: 10.1201/9781003396949-11

businesses come in all shapes and sizes from local businesses to national and transnational corporations, branch plants, sole proprietors and partnerships, not-for-profit, and even state-owned enterprises. Investors are occupied by management thinking, meaning a dependence on good governance, strong leadership, strategic thinking, and sound management practices. While firms themselves are generally short-term oriented, hoping to achieve rapid returns on capital, investors will take a long view, focusing on reliable returns over an extended period. In addition to the condition and performance of the physical asset, investors will perform due diligence on the people involved, their skills, the culture they represent, and the stability of structure within which they operate. Investors will want to know how the organization may fare in the face of uncertainty, complications within the market sector, and external threats. They will be interested in market conditions, competitors in the market, and pricing relative to the competition. This is not to imply that they overlook ways to enhance the social welfare of workers or fail to appreciate the responsibilities to address issues dealing with the social and economic well-being of society. Such failings only increase risks that are hard to price.

Private investors have a responsibility to achieve a return-on-investment on their capital, and infrastructure is just one of many investment opportunities available to them. They are not as committed to infrastructure as they are committed to meeting their investment criteria from two perspectives: whether a particular investment meets threshold requirements; and how the investment contributes to their overall portfolio of investments. By buying operating businesses investors are buying a "guarantee" of future profits over which they have control and hopefully can build a platform upon which to grow their investment. By buying businesses they are not just acquiring an asset, but also acquiring skilled people to manage and operate the asset. In buying businesses, private investors focus on management thinking.

The Financial Crisis of 2008 marked a turning point in the involvement of government and the private sector in the provision of infrastructure (Bryson et al., 2014). For governments, it meant increasing austerity in the delivery of public services and cutbacks on "business-as-usual" approaches to providing. At the same time, private investors in infrastructure were confronting an unusual challenge—too much capital versus too few investment opportunities.

To deploy this capital and address market opportunities, investors realized that new and innovative business models were needed with the potential to bridge the interests of both the public and private sectors. No one was sure what these models would look like, although precedents and prototypes have subsequently emerged. New business models will be required for both brownfield and greenfield infrastructure, with an understanding by governments that private investors are not bankers. As owners, they are primary sources of equity in transactions that can maximize both absolute returns and net value-added returns. They are also risk takers for which they want to be adequately compensated. Investment decisions will continue to be made within a broad framework of overall investment market choices, but with an increasing interest in elevating infrastructure to one of their preferred choices.

Simplifying the business proposition

Investors rely upon a portfolio approach, depending on three basic sources to contribute to not only absolute returns, but net value-added returns. Achieving targeted portfolio returns is not a simple exercise as Figure 8.1 may imply. Success depends upon a complex series of management decisions from beginning to end under the rubric of enterprise risk management (Bryson et al., 2014).

Enterprise risk management

Enterprise risk management (ERM) is a top-down strategic decision-making process directed at identifying and preparing for hazards that can potentially affect a company's finances, operations, and objectives. ERM allows managers to shape the firm's overall risk position by mandating certain business segments engage with, or disengage from, particular investment choices. Each organization will have its own risk management culture based on the existing risk profile of the organization, as well as the new risks it is willing to take on. This will be materially different across investors, although all will share a need to understand the nature of the risks they are contemplating, how to measure these risks, and what mitigation measures will be required. It is the risk enterprise exercise that directly impacts target asset mix and capital allocations to the various asset classes. Risks can be assigned under five categories[1]:

- **Governance risk:** everything from audit and oversight of financial systems to conformance with legal and reporting requirements and adherence to regulatory regimes. Most of these are responsibilities of a board and reflect the competency of board members and the effectiveness of the board in carrying out its governance functions.
- **Strategic risk:** identifying, assessing, and managing risks that could inhibit an organization's ability to achieve its strategic objectives that can arise from both internal and external events (Frigo and Anderson, 2011). The goal is

Figure 8.1 The three sources of total portfolio return.

Source: CCP Investments. 2021 Annual Report.

to achieve and protect value. Strategic risk management is viewed as a core competency of senior management and the board.

- **Investment risk:** during the investment strategy formulation, a greater focus on expected returns and lesser on risk may result in a mismatch between the investment objective and risk appetite. The objective of investment risk management is to provide assurance that assets are being managed according to stated investment strategy, consistent with the portfolio object-ives (as defined by the investment mandates), and according to professional standards of monitoring, control, and accountability (Jaggi, 2018). Each organization's risk management culture, size, scope, and geography will give rise to different practices to determine an appropriate level of control of investment functions.
- **Operational risk:** risk of loss resulting from ineffective or failed internal processes, human error, system failures, or external events that can dis-rupt the flow of business operations. The losses can have direct or indirect financial consequences (AUDITBOARD, 2018). Operational risk can refer to both operational risks within an organization and the processes man-agement uses when implementing, training, and enforcing policies. While operational risk management is considered a subset of enterprise risk man-agement, it excludes strategic, reputational, and financial risk.
- **Reputational risk:** threat or danger to the good name or standing of a business or entity[2]. As one leading CEO stated the case, "Do no wrong". Reputational risk can occur directly, as the result of the actions of the com-pany; indirectly, due to the actions of an employee or employees; or tan-gentially, through other peripheral parties, such as joint venture partners or suppliers. In addition to having good governance practices and transpar-ency, companies need to be socially responsible and environmentally con-scious to avoid or minimize reputational risk.

A portfolio approach

Investing in an infrastructure asset is a relative, and not an absolute, decision. According to CPP Investments, investment decisions are made within a context of an overall portfolio return derived from a combination of three factors: diver-sification, investment selection, and strategic positioning (Figure 8.1)[3].

1 **Diversification** is the most powerful instrument to mitigate market downturns and enhance long-term investment returns without increasing portfolio risk. This means investing in fundamentally distinct sources of value creation that can combine exposure to both systemic and non-systemic risk. Infrastructure has strong diversification qualities and can be acquired through three types of investment, identified below. In the case of the public and private equity investments, these might not be classified as investments in infrastructure for the purpose of portfolio construction, and may be assigned to public equity, private equity, and credit.

- **Public market investments**. These investments can capture global economic growth through equity ownership and credit investments in public traded companies around the world. Examples could include shares in listed energy, water, or telecommunications companies.
- **Private company investments**. This includes equity and debt in privately held companies, both direct and through funds and partnerships. Returns are generated through corporate earnings and dividends.
- **Real assets**. These assets include real estate and infrastructure and generate returns from various sources such as property income, facility user fees, and income from renewable energy contracts or the sale of natural resources. Higher returns are expected to compensate for additional risks, complexity, and illiquidity. These assets are typically held through private corporations, partnerships, or other entities. Examples in infrastructure could include airports, alternative energy production and distribution, highways, water and wastewater, refuse, and ports.

Figure 8.2 depicts the asset mix of OTPP in June 2020, showing the above categories in a portfolio context. Infrastructure and real estate represent real assets. This portfolio is designed to exceed the 4 percent real return hurdle needed to pay pensions based on actuarial forecasts.

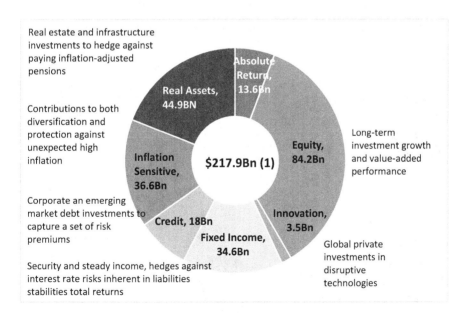

Figure 8.2 Ontario Teachers' Pension Plan (OTPP) Investment Portfolio, 2020.

(1) Assets net of investment-related liabilities.

Source: Ontario Teachers' Pension Plan, 2020.

2　**Investment selection.** Each investor will have their own approach to identifying, selecting, weighting, acquiring, and disposing of investments. This is
  where skilled internal and external investment managers make a difference
  when it comes to outperforming the market. This is not an easy task in
  view of stiff global competition. An advantage of the large institutional
  investors using a direct approach is the avoidance of fees, the benefits of
  scale, their long-term investment horizons, and the ability to become more
  "efficient" by capturing and retaining market knowledge within their own
  professional teams.
3　**Strategic positioning.** Strategic positioning is a deliberate and meaningful,
  but perhaps a temporary shift of asset allocations and/or exposures away
  from a portfolio's established targets or normal mix[4]. An example would be
  responding to disruptions in the market from COVID. Strategic positioning
  can be used to exploit gaps in the market, address market volatility, take
  advantage of declining prices in certain sectors, or address other stress
  conditions. While not a replacement for diversification, strategic positioning
  can, from time to time, add to total returns, protect asset values, and fund
  liquidity.

Figure 8.3 is a snapshot of the consequences of a dynamic and ever-changing
decision-making process. This depiction places the investment portfolio

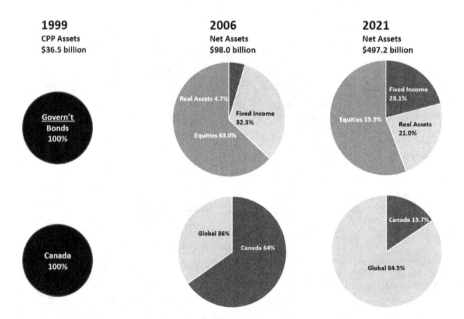

Figure 8.3　The Evolution of CPP Investment Strategy, 1999–2021.

Source: CPP Investments. Annual report 2021.

in the context of time and reflects the evolution of CPP Investment's strategy from 1999 through 2021. A key feature of this evolution is the heavy weighting of global investments, 84 percent of the total portfolio, as compared to 100 percent Canadian when the fund was launched three decades ago.

Portfolio construction

BlackRock defines portfolio construction as "the process of understanding how different asset classes, funds and weightings impact each other, their performance and risk and how decisions ladder up to an investor's objective"[5]. They view portfolio construction as a four-step process:

- **Benchmarking** involves establishing a standard point of reference against which to measure portfolio performance and compare and evaluate asset allocation choices. Various benchmarks may be used based on the investment approach, ranging from measurements relative to a total investible universe to measurement relative to a specified outcome.
- **Budgeting** refers to two types of budgets—cost and risk. The cost budget is directed at minimizing the costs of fees, taxes, and other expenses to maximize the value of the portfolio. The risk budget sets the level of risk based on investment objectives. This budget defines the acceptable risks level and what types of risk an investor is willing to take on.
- **Investing** is governed by the investment objectives of the portfolio, the due diligence process that any investment is subject to, and determination of the impact of the investment on overall risk-adjusted returns. This impact can either enhance return or reduce risks.
- **Monitoring** is a discipline applied on a frequent basis (monthly or quarterly) to determine whether the investment strategy requires adjustments. Adjustment may respond to changing investment objectives, evidence of unintentional risks in the portfolio, and changing market conditions. Rebalancing may have clear long-term benefits but also raise some short-term concerns with investors.

A good example of a rapidly changing investment landscape is the dilemma that faced investors in airports across the world that all but shut down in March 2020, with the exception of cargo, in response to COVID (Bouwer et al., 2022). Large institutional investors involved in major international airports chose to ride out the situation in expectation that their long-term prospects in this sector remained sound. That is proving to be a wise decision as air traffic has returned to pre-COVID levels, but not without some serious disruptions in airport operations. According to the McKinsey report, during the COVID lockdown, air traffic fell by as much as 95 percent, with the airlines taking the biggest hit. Of a total loss in 2020 of $230.1 billion, 73 percent of the loss is attributed to the airlines, with the next largest loss being the airport themselves at 14 percent of the total loss.

Structuring an investment business

Business decisions are shaped by a top-down process, driven by the risk appetite of an organization and its fiduciary responsibilities. This risk profile is the level of risk that would be expected to generate the net real return required to sustain the obligations of the investment fund. A framework for these decisions is depicted in Figure 8.4, representing the obligations of CPP Investments. The source of funds is the payroll deductions that workers in Canada, and their employers, jointly contribute to the Canada Pension Plan. Priority on these contributions is used to pay CPP benefits, and the remaining funds not needed to pay benefits are used to create a global investment portfolio. This is how CPP Investments grows the value of the total fund to meet obligations to pension holders that would not otherwise be covered by incoming contributions (Figure 8.5).

Factors influencing a portfolio investment approach

A strategy for investing in infrastructure is typically predicated on several factors that define a total portfolio approach for most infrastructure investors.

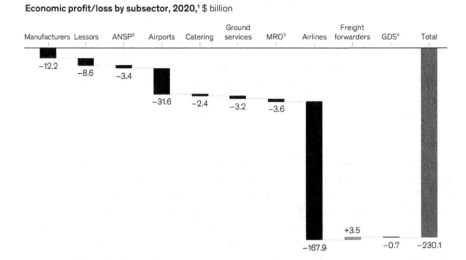

Figure 8.4 Value destroyed in the aviation sector in the first year of the pandemic.

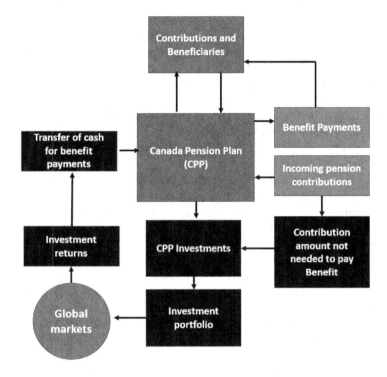

Figure 8.5 CPP investment framework.

Source: CPP Investments. Annual report 2021.

These factors will vary across funds, and collectively, will define the competitive advantage of each fund.

- **Investment horizons**. For the large institutional investors, long payoff investments reflect the obligations of a fund to pay pensions and in the case of insurance companies pay out insurance benefits. This may not be as true for the unlisted funds that could take a short-term approach in response to business pressures, short-term market fluctuations, or regulatory considerations.
- **Cash flow requirements**. Institutional investors will want a predictable and stable pattern of cash flows over time. They will avoid situations where they must sell assets to pay benefits. Investments will be based on reliable forecasts of future investment needs. With unlisted funds, their cash flow requirements will depend upon the structure and obligations of each closed-end fund.

- **Scale**. This is what sets the large institutional investors apart. They can seek opportunities globally in private markets or engage in public market strategies that exceed the capability of smaller managed funds. A large-scale operation can support in-house technical teams, provide confidence to local partners, introduce new support technologies, cover the costs of market research, and build long-term investment platforms in new markets.
- **Brand recognition.** The large institutional investors are brand conscious, as a brand is a proxy for reputation. The same applies to the managers of the unlisted funds as their fundraising activities are built upon reputation and trust that offer significant competitive advantages in hotly contested markets.
- **Partnerships**. Being a global business involving very large amounts of capital, operating across multiple continents and in various infrastructure sectors, partnerships are an essential part of any investment strategy. Partnerships are not only a risk mitigation strategy, they provide access to new investment opportunities, local market knowledge and insight, and are essential participants in dealing with turbulent or crisis situations. Partners can offer particular expertise and may deliver asset management services.
- **Risk appetite**. Investment returns cannot be earned without taking on risk. No two funds are the same in terms of their respective risk appetites, and this applies to both the large institutional investors and the unlisted funds. Risk appetite drives strategic decision-making and is the single most important factor in building a long-term investment portfolio. The level of risk correlates with return expectations—the higher the risk target, the higher the long-term expected returns. However, higher risk also implies exposure to short-term volatility and the possibility of increased losses.
- **Management**. Large institutional funds are actively managed in-house. This is not a low-cost approach. In-house management increases the complexity of the organization, requires additional resources, and must be justified in terms of effective risk management and increased returns. It also requires a large organization to be able to construct a portfolio that is significantly more diversified, by asset type, region, and sector, and includes private equity and real assets. Effective management can be a significant competitive advantage.

Governments talk about attracting private capital to fund infrastructure, but this desire may not align with realty and what investors are looking for to fulfill their mandates. Investors have a private equity mindset with a strong business focus. For these investors, this means acquiring and building companies with the appropriate business model, the right risk assessment procedures, and the right people in the right places to grow the business.

Notes

1 Industry Day Presentation. OTPP. Sustainable Infrastructure Fellowship Program. Slide 6.
2 www.investopedia.com/terms/r/reputational-risk.asp
3 CCP Investments. 2021 Annual Report. P28-29
4 CCP Investments. 2021 Annual Report. Ibid P29
5 Sourced at www.blackrock.com/americas-offshore/en/education/portfolio-construct ion/understanding-portfolio-construction#:~:text=What%20is%20portfolio%20c onstruction%3F,up%20to%20an%20investor's%20objectives

9 Elements of the business model

It is not possible to discuss investment of private capital in infrastructure without considering the business model through which this capital is invested. Even with debt, a business model is a key piece of the puzzle to ensure that the debt can be retired. For institutional investors, the preference is for business models in a mature state of operation. Business models have always been at the core of any appeal to engage the private sector as evidenced by the many historical precedents that shape the infrastructure we rely upon today.

During the 19th century, new infrastructure technologies such as railways, telegraph, telephony, and electricity systems were introduced in the U.S., and to a lesser extent in Europe, by competing private firms (Davies et al., 2010). Perhaps as important as the technology were the business models to capture customers, generate revenues, and enhance value. AT&T developed a business model based on telephone infrastructure to provide customers with "one system, one policy, universal service" under the control of a large, vertically integrated, and centrally managed bureaucracy. Government soon realized that infrastructures in transport, water, energy, and communications are "natural monopolies" that may require public ownership, as in Europe, or government regulation as in the U.S. (Davies et al., 2010).

AT&T avoided intense pressure for public ownership, as had occurred in Europe, by offering cross-subsidized telephone service to urban and remote rural communities under Federal regulation. But the private sector increasingly found it could not compete in a framework of heavy regulation, or with public monopolies that were created by governments to prevent market failure and meet universal service obligations. The provision of essential public infrastructure began the shift from the private sector to public entities, controlled and paid for by state and local governments. This trend continued throughout the early decades of the 20th century with exceptions in a few sectors. Railroads, shipping, and energy transmission including power grids and pipelines, largely remained private, all of which had very large capital requirements, were technically and operationally complex, and spanned multiple political jurisdictions.

Thatcherism, starting back in the early 1980s, brought an end to always relying on the public sector with its emphasis on free markets, reduced

DOI: 10.1201/9781003396949-12

government spending, tax cuts, deregulation, and privatization. It had a counterpart in the U.S. with Reaganomics. Thatcherism directly influenced public policy decisions in Australia, New Zealand, and Canada and spawned a worldwide economic liberal movement. The shift from a regulative, bureaucratic style of government towards a strategic, market-driven governance style was embodied in the principle of New Public Management (NPM) (Christensen and Laegreid, 2011). NPM refers to the reshaping of the public sector brought about by privatization, the restructuring of public services, and introduction of private market disciplines into public administration.

According to NPM, the injection of market pressures and private sector skills would improve the quality of public sector services. This did not bode well for public programs heavily vested in delivering public infrastructure, including many state-owned enterprises and monopolies in sectors such as telecoms, energy, roads, water, and airports. Some of these were privatized, such as airports and ports, and others were broken up and split between public and private ownership, such as water and energy. New competition in procurement was introduced with private finance initiatives (PFIs) in the UK, public–private partnerships (PPPs) in Australia, and alternative finance and procurement (AFP) in Canada, all derivates of PPPs[1]. PPPs presented a new interface between the public and private sectors with intended public benefits that are still questioned today.

A manifestation of Thatcherism was reference within governments to the term "business model". Governments began to adopt terms and definitions from business, not quite understanding their relevance in a social contract context. They talked of new business models with the capability to attract private know-how, private capital, more efficient management practices, technical and process innovations, and effect cost savings for taxpayers. A harsh critic of this shift was McGill University business professor Henry Mintzberg, an acclaimed management thinker, author, and iconoclast. In his widely read article in the Harvard Business Review, 1996, he stated that, "Above all, say many experts, government must look like business. It is especially this notion I wish to contest" (Mintzberg, 1996).

Unfortunately, the drive to engage the private sector in public infrastructure did not move the needle much beyond PPPs. Nor did PPPs translate into business models as understood by the private sector. Time has shown that PPPs garnered very limited private participation beyond taking on long-term operational responsibilities and achieving net returns on small equity investments in the range of commercial bonds. For the private sector, PPPs are not a framework that drives a motivation to be profitable, nor do they deal with the internal logic of how a private partner creates and captures value.

Private firms look for ways to organize value-adding activities and structure transactional relationships with clients, customers, suppliers, market competitors, and stakeholders. Returning to Henry Mintzberg once again, in another of his well-known articles in Harvard Business Review, Crafting Strategy, 1987, and based on his empirical research, Mintzberg claimed that "managers are crafting, and strategy is their clay" (Mintzberg, 1987). The implication being that a manager, much like the craftsman shaping a lump of clay on the potter's wheel, merges formulation and implementation in a fluid process to achieve desired results, the outcome of which is not easy to predict or predetermine. This notion of "crafting" is the opposite of what a PPP involves and highlights the difference between an entrepreneurial business venture and a set of legally binding agreements supposedly guaranteeing a prescribed outcome.

Following a seismic shift in political ideology under Thatcherism in the early 1980s, a turning point in the search for increased engagement of the private sector was the financial crisis of 2008. This is when governments realized they no longer had the financial means to support traditional approaches to procuring and operating infrastructure. Governments had financed public infrastructure through various forms of taxation, but austerity politics put an end to this. The private sector now had an opportunity to step up its role in developing, funding, and maintaining infrastructure. The unanswered question was whether new business models could be launched in time to accommodate large upfront capital investments with long-term stable revenue streams, and still be acknowledged by government as serving the public interest.

The difficulty for both sides in this discussion is that no one quite knows what a business model is, least of all governments. There is no single business model, nor convenient diagram such as with PPPs, that can capture what a business model is. Business literature contains various definitions and explanations, with a heavy emphasis on innovations that can transform existing markets. These depictions are usually ambiguous and imprecise and not particularly helpful. Much of the recent academic literature is preoccupied with examples of highly successful private firms that created innovative business models suited to intermediating in existing markets.

From the 1990s onward, the term "business model" in the private sector has been used to describe innovative ways of practicing or doing business with an emphasis on information and communications-based technologies, the Internet, and e-commerce. It is interesting to look at definitions offered by the experts. These definitions are all different but share an underlying theme: how to make money (Bryson, 2014). Recent examples of new business models that have brought about radical change or even disruption in the marketplace reflect new technologies, the Internet, the burgeoning world of data collection, and the sharing economy. Some of the well-known names include Uber, Shopify, Spotify, Airbnb, Facebook, Netflix, Tesla, and Skip-the-Dishes. Beyond the big names are new business models quietly dominating industry sectors from retail and banking to travel, gambling, health care, and personal services.

When it comes to Amazon, Apple, and Microsoft, it is impossible to succinctly summarize their business models, and even more challenging to comprehend the sources of revenue and resulting profits that drive these businesses. They have never faced a severe economic downturn, until now—they are currently shedding employees in a reversal of trend. Determining value adds to this confusion, particularly when market capitalization reaches astronomical amounts, despite poor returns. Many attempts at business model innovation fail, and according to Roland Christensen at the Harvard School of Business, success appears to be random (Christensen et al., 2016). Christensen claims that confusion around business models starts with confusion around what the term "business model" actually means.

Elements of a business model

The business guru, Peter Drucker, claims that the purpose of a business is to create a customer, a comment which offers a starting point for any discussion on business models (Drucker, 1954). The basis of this comment is the underlying search for a meaningful value proposition which can define a product or service offering. The value proposition is one of four interrelated elements that Christensen and his colleagues identify as the basic elements of a business model (Christensen et al., 2016).

- **Value proposition**: A product or service that serves customers in doing a job they require, is effective, convenient, and affordable.
- **Profit formulation**: This includes assets and fixed cost structures, and the margins and velocity of the revenue stream to recover these costs and generate a profit.
- **Resources:** People, technology, other products, facilities, equipment, and capital required to sustain the value proposition and satisfy customer demands.
- **Processes:** How the various components and parts work together and interact in a consistent manner to affect the value proposition.

The interaction among all four, over time, serves two purposes. First, it applies a framework to understand and assess the capabilities of a business; second, it highlights the interdependencies among the key elements of a business and reveals what an organization is capable of achieving, plus understanding what it cannot do. However, this is not a static situation and time plays a huge role in influencing performance ranging from the cost of money and changing markets, to the results of ongoing research and development.

According to Christensen, a business goes through three stages as it evolves over time. Stage One is "creation", when the business establishes and then refines a value proposition. In this stage, a business will operate with minimal resources in its search for insights that can shape its recurring tasks. Stage Two is "sustaining innovation", and this occurs when customer demand is proven, and the challenge is rapidly scaling an operation up to meet this growing demand. This stage involves the building process and searching for better products at higher prices to enhance the value proposition and meet customer expectations. Stage Three is the "efficiency" stage and the search to reduce costs that can involve everything from outsourcing to adding financial or operating leverage, optimizing processes, and consolidating to gain economies of scale.

Christensen highlights the fluidity and flexibility required of a business over time as it responds to changing needs and demands. Business models are predicated on the fact that priorities constantly shift to capture and retain value. An organization 5–10 years hence may look very different from what was envisioned at the time of creation. This is the opposite to what is expected of a PPP. With a PPP, the asset must be returned to the legal owner at the end of the concession period in essentially the same condition as when acquired.

Any infrastructure business that delivers a product or service will take a long-term view given the upfront capital required, investment objectives, and the challenges of acquiring these businesses in a very competitive environment. This is what distinguishes greenfield from brownfield operations. Greenfield implies a heavy emphasis on the "creation" stage of a business, with many unfamiliar or unknown business risks. Brownfield investments might be in the stage of "sustaining innovation", but preferably in the "efficiency" stage where the value proposition is proven and there is still opportunity for value creation.

Public–private interface

Difficulties arise when attempting to translate business model thinking into public sector markets. Private markets thrive on competition, customer satisfaction, consumer choices, value-added products and services, market efficiency, and continuous improvement through research and development. Private organizations only survive in the long run if profitable. In comparison, public markets are highly regulated, seldom face direct competition, may not distinguish among customers with different needs, have few incentives to introduce technological advancements, and have pricing that is often distorted by subsidies and grants. A public entity may be required to deliver a service whether it generates revenues or not.

There are constant calls by governments across the globe to attract private capital to address infrastructure needs, including increasing private investment in PPPs. In Canada, the mandate of the recently founded Canadian Infrastructure Bank (CIB) is "… . to invest $35 billion in revenue-generating infrastructure which benefits Canadians and attracts private capital"[2]. What these pleas seldom reference is the parallel need for new business models to structure these investments.

The Local Government Association in the UK, in a January 2022 press release, stated that "PPPs can be well suited to addressing housing and regeneration challenges. They can ensure these strategically important projects are delivered and accelerate delivery through availability of capital. In some cases, the majority of the risk and capital can be borne by the private sector"[3]. In the emerging economies, and despite a similar plea to attract private investors, according to the Global Infrastructure Hub private investment in infrastructure has remained stagnant for the past seven years, and lower than it was 10 years ago[4].

In both developed and developing economies, the challenge for increasing the participation of the private sector in new infrastructure projects is fundamentally the same: How can new business models deliver infrastructure projects in an environment dominated by public sector organizations that are heavily committed to traditional EPC procurement or PPPs as a preferred or even exclusive procurement process? Difficulties arise when it comes to translating the basis of a preferred business model to regulated or publicly owned infrastructure providers. There is also a misalignment between the public and private sectors when it comes to understanding a value proposition, determining the scope of the enterprise, fostering transactional relationships that extend from management to users and suppliers, and the role that people play in translating a business model into reality. Until a better understanding is reached between the two sides, the private sector will continue to redirect its money into secondary deals for which they can identify risks they are willing to accept, and match these against their own objectives.

According to Davies, a public–private interface will require an understanding by the public sector of six main elements or components that serve as a foundation upon which a business model can be constructed (Davies et al., 2010). This involves a government adapting to a way of thinking, being familiar with business terminology and definitions, and understanding the core logic driving a value proposition. One can assume governments have not previously encountered most of these challenges and have limited understanding of their implications. The six elements that Davis refers to are:

- **Strategy.** A business model is based on a core strategy. Strategy can be seen as a multidimensional concept that embraces all the critical activities of the firm, providing it with a sense of unity, direction, and purpose, as well as responding to the environment in which it will operate (Hax and Majluf, 1996). Strategy serves as a unifying point of view, a fundamental framework through which an organization can assert its vital continuity and adapt to a changing environment.
- **Value proposition.** A business model will be predicated on a value proposition that will set out the operational tasks to achieve the strategic objectives. The value proposition identifies the market segments that are targeted for profitable growth and sets out the potential profitability of the business model.
- **Value chain and internal capabilities.** A value chain highlights internal capabilities, resources, communication/decision channels, consumer knowledge, strategic assets, and the competitive advantage of the firm. A business model will identify which part, or parts, of the value chain the strategy is targeting. Think of IKEA with a business model that involves a retail outlet configured both as a showroom and a warehouse and products that the purchaser picks up as boxed parts that they must assemble themselves.

- **Organizational form.** A key element of a business strategy is how a firm is organized to address the opportunity. A vertically integrated firm can be responsible for the entire value chain, whereas some organizations must rely on extensive outsourcing and partner relationships. Organizational form will reveal the extent to which new technologies can be incorporated and new customer offerings can be introduced. Compare the software model that Google applies versus that of Apple. Google's Android is considered an Open-Source mobile, while Apple's iOS is considered closed source. Google's open model gives developers the option to alter things as they wish. Apple does the opposite.
- **Value network.** As distinct from the value chain that represents activities performed within the firm, the value network represents the entire network, internal and external, of clients, suppliers, partners, and other key stakeholders involved in delivering value to the customer or end user. Where a firm clearly positions itself in the value network is a key determinant of the eventual success of a business model.
- **Value capture.** This refers to how an investment is monetized and includes the underlying cost structure, revenue sources, profit formulas, and how risks and rewards are shared between the firm and its suppliers and partners. Value capture is a long-term proposition that should be scalable, repeatable, and efficient.

These six factors highlight the competitive advantage of institutional investors who have the capacity and ability to acquire infrastructure assets directly and manage these assets internally. Compared to the unlisted funds, the institutional funds are more inclusive of all facets of their operation, their value chain is usually more extensive, and they can approach value creation in a much broader context. But this comes with added risks and significant imbedded costs that must be accounted for.

These six factors highlight the dilemma of funds involved in brownfield acquisitions and seeking to extend their involvement into greenfield projects. Greenfield projects may fall well beyond the scope of their internal capacity, their value chains, and even their organizational structure. Investors seldom have a comparative advantage in a market that requires them to first make something to invest in, particularly when this process includes active participation in, or regulation by, a government entity. This reticence is reflected in the low rate of involvement of institutional funds in PPPs. These challenges can be overcome in some circumstances by setting up wholly owned subsidiaries specific to greenfield projects as some pension funds have done[5].

Business models for greenfield projects

Most greenfield infrastructure projects involving a public–private interface are one-off projects that generally fall into four categories (Davies et al., 2010):

1 **Risk transfer model.** A prime contractor is responsible for managing the project and this is typical of most large PPP projects under a design–build

arrangement, as well as smaller capital projects. It is difficult to characterize these as business models as they are essential legal contracts embodying risk assignment and revenue allocation in which the investor is relegated to a passive role in support of a special purpose vehicle (SPV).

2 **Risk negotiation model.** Small infrastructure capital projects are exploring other options to managing risk including integrated project delivery (IPD) developed in the U.S., and alliance contracting, developed in Australia, both representing collaborative approaches (Australian Government, 2015). The Thames Tideway Tunnel is an example of the alliance approach[6]. Risk allocation is not pre-ordinated, but rather is the result of extensive upfront public negotiations over the allocation of risk.

3 **Risk sharing model.** This involves a partnership and risk sharing, based on extensive dialogue at the outset of a project and trust among the partners. The model depends upon a new cooperative and semi-permanent (e.g., five years) organizational arrangement including integrated project teams. An example is the Urban Partnership Strategy involving Yorkshire Water, ARUP, and Costain. The ARUP/Costain JV provided expertise in construction, design and management regulation compliance, and design phase process as well as construction site safety advice to help Yorkshire Water address affordability challenges in the delivery of its five-year AMP7 investment plan[7].

4 **Risk bearing model.** This model was developed by BAA to design and build the T5 terminal at Heathrow Airport. The client retained responsibility for the risk of cost and time overruns, and failure to achieve the desired outcomes[8]. This required an extensive in-house capability to preside over all strategic decisions and act as the systems integrator to coordinate and control the project. This model is supported by cost-reimbursable contracts and a partnering approach with contractors to reward cooperative efforts to expose and manage risk, rather than mitigate risks that arise.

Expanding involvement by institutional funds into greenfield infrastructure projects will require four things:

1 A business model that embodies their core strategies.
2 A defined revenue stream, combined with value-added opportunities to drive a value capture proposition.
3 An internal resilience that can support fine tuning of the model, increase the capacity for innovation, and respond to changing market conditions or new requirements such as sustainability targets.
4 An asset that be sold or traded, in part or in whole, and at any time.

The third requirement is something that governments will be uncomfortable with as they seek certainty in their arrangements, rely on legal binding contracts, with enforceable penalties if performance fails to meet expectations. The fourth requirement provides the flexibility to liquidate the asset, bring in partners, or facilitate withdrawal of some of their equity at a future time.

Government partners are often hesitant to support the subsequent sale or transfer of carried interests as it may be construed in the media as a profit grab by the private partner.

Business models for emerging economies

The work of C.K. Prahalad is recognized as pioneering in setting a conceptual framework for business models best suited to the constraints of emerging economies. His seminal work, written with Stuart L. Hart, "The Fortune at the Bottom of the Pyramid", was published in 2002 (Prahalad and Hart, 2002). The book addresses the business opportunities at the "Bottom of the Pyramid" (BoP), as illustrated in Figure 9.1 which updates Prahalad's calculations and only reinforces his claim.

Prahalad popularized the idea that the lowest cohort represented a profitable consumer base given appropriate business models. His thesis is not without its critics and detractors, particularly among fellow academics. However, there are examples in India, South Asia, and Africa where successful enterprises have been built around Prahalad's bottom-of-the-pyramid thinking. These include the microcredit market in South Asia, health care, mobile phones and mobile

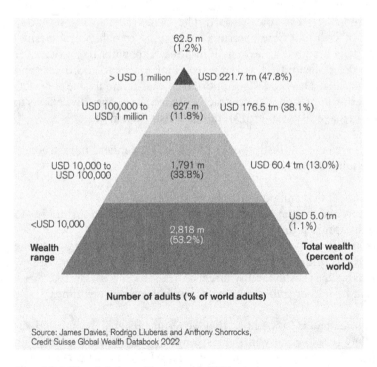

Source: James Davies, Rodrigo Lluberas and Anthony Shorrocks, Credit Suisse Global Wealth Databook 2022

Figure 9.1 The global wealth pyramid, 2021.

banking, waste management and disposal, and household cleaning products geared to slum settlements in India.

His thinking was predicated on reinterpreting the framework of profitability used by businesses across the globe. Given that the ability to pay by those at the bottom of the pyramid could be as little as 10 percent of what consumers in the upper levels of the pyramid can afford to pay, it was Prahalad's thesis that traditional business approaches to profitability will not work for a large segment of the world's population.

Traditional thinking is:

PRICE = COST + PROFIT (cost structure is a given and profit is the motivating factor—this is the dominant logic of capitalism). Prahalad proposes a different formula.

PRICE – PROFIT = COST. Challenging the dominant logic, Prahalad posited price is what people can afford, profit is what motivates, and cost is based upon the capacity to consume.

Prahalad then laid out his "12 Principles for BoP Innovation"[9]. This emphasis on cost drove the search for innovation to drive down costs such that the final price fell within the range of what consumers at the bottom of the pyramid could afford. A well-cited example of Prahalad's thinking is India's Aravind Eye Care System, the world's largest eye-care business that performs hundreds of thousands of cataract surgeries each year[10].

Aravind, with its mission to "eliminate needless blindness", provides large-volume, high-quality and affordable care. Around 50 percent of its patients receive services either free of cost or at steeply subsidized rate, yet the organization remains financially self-sustainable. Much importance is given to equity—ensuring that all patients are accorded the same high-quality care and service, regardless of their economic status. A critical component of Aravind's model is the high patient volume, which brings with it the benefits of economies of scale. Aravind's unique assembly-line approach increases productivity tenfold. Over 450,000 eye surgeries or procedures are performed a year at Aravind, making it the largest eye care provider in the world.

In developed markets, Prahalad suggests that one cannot assume the availability and reliability of electricity, telephones, credit, refrigeration, and other such amenities. At the BoP, the infrastructure is much spottier and more hostile. Consumers may have to cope with frequent electric-power blackouts and brownouts. Credit may be extremely costly. Refrigeration may be unavailable. Products marketed to the bottom of the pyramid must be able to withstand fragile environments.

A recent example of Prahalad's thinking, applied to an essential infrastructure component, is that of Mukesh Ambani's approach to the low-end cell phone market in India. Priced around $50 and supported by a low-cost wireless plan from Reliance, an Ambani company, the JioPhone is an Android device custom built for the India market by Alphabet Inc., a subsidiary of Google

(Bloomberg, 2021). The plan is to sell 200 million "JioPhones" over the next two years. This device can be used for banking, paying rent and utility bills, loans, and receiving payments. Google gets to access the data and Facebook gets to promote it social media platforms and text messaging apps.

The relevance of Prahalad's work to investing in infrastructure in emerging economies is fourfold. First, an emphasis on innovation. Solving problems in the emerging economies cannot be based on versions of developed-world products. Products must be rethought to bring radically lower cost while retaining high quality and good value. Second, solutions must aim to conserve resources, and this places an emphasis on environmental sustainability. Innovations must emphasize conserving resources, recycling materials, and eliminating waste. Third, the design of products and services must be capable of enduring tough infrastructure environments where reliability is not possible and skilled labor may not exist. This places an emphasis on process innovation to deal with interruptions and lower expectations on functionality. "Good enough" is an acceptable mantra if cost is minimized. Finally, traditional assumptions must be challenged and conventional wisdom cast aside when it comes to setting prices and what may be possible at a substantially lower cost.

Infrastructure business models

Academic literature on business models largely ignores developments in the public–private sector, and yet there are significant advances in developing business models that apply to regulated and/or publicly owned infrastructure projects. So much of the available literature on business models dwells on the firm itself, and yet fertile ground for new business models is understanding how clients and partners can work in new collaborative environments to create and capture value in one-off projects, or semi-permanent arrangements (Davies et al., 2010). These arrangements may involve changing the balance of responsibilities, risk assignments, and relationships between the public and private domains, and the introduction of innovations or new organizational structures.

A new frontier for business model innovation will be responding to the demands for sustainable infrastructure. The development of new large-scale urban infrastructure projects, from clean energy generation to resiliency measures, will require new business models and financing instruments. Today, governments are utilizing an array of sources, including green municipal bonds, sustainable revenue streams, tax incentives, private sector sources, credit enhancements, and international or crowdfunded support, among others. These sources of capital, in addition to private capital, open new business opportunities.

These opportunities fall into two general categories as depicted in Figure 9.2, expansion into existing businesses and diversification into new businesses (Hax and Majluf, 1996). For most investors, expansion into existing businesses, if the opportunities exist, will be the preferred route as this minimizes risk. However, this will be a strategic decision that an organization must make in light of

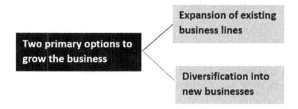

Figure 9.2 Alternatives for growth and diversification.

Source: Hax and Majluf.

factors that range from micro- and macro-economic conditions, growth trends, technology, internal strengths and weaknesses, to the vision of the firm. These are decisions deeply rooted in the corporate culture of an organization.

Growth will be the dominant force driving the search for new business opportunities (Hax and Majluf, 1996). For infrastructure investors, the primary strategy will be international expansion of current business lines, within the current structure of the organization. A secondary strategy available to investors will be vertical integration, attempting to increase value-added within a given business base. This will happen in two directions: forward, which leads an organization to be closer to its customers/clients; and backward, which moves it closer to its suppliers.

The logical next step is to seek entry into new businesses via diversification. Related diversification is supported by the expertise residing in one or more stages in the value chain. Unrelated diversification is referred to as conglomeration. In either case, growth can be achieved by internal development or acquisitions. Internal development has the advantage of using existing resources and cultural consistency, whereas acquisitions are speedy and take care of a lack of skills and competency within the organization. The challenge in all cases is the choice of an appropriate strategy. This implies a careful analysis of the growth opportunities available to the organization as the basis for selecting a preferred strategy.

An example of a successful growth strategy in the infrastructure sector is that of Brookfield, a firm with origins that date back more than a century. The Brookfield universe is not easy to comprehend, even to those in the industry given the diversity and complexity of their holdings (Frampton, 2023). They have operating businesses in infrastructure, renewables, real estate, private equity, and credit, plus insurance operations. They have multiple sources of revenue generation and in addition to managing their own money, they manage capital for other clients on a fee basis. They generate profits from insurance premiums, cash flow from operating businesses, management fees from pension funds, endowments, sovereign wealth funds, and insurance companies. Brookfield also generate capital from the sale of their own assets and redeploy this capital to acquisitions where there are better prospects for higher returns.

Notes

1 Governments in both the UK and Ontario, Canada, chose not to include the words "private" and "partnership" in their terminology for PPPs, strictly for political reasons.
2 Sourced at https://cib-bic.ca/en/about-us/our-purpose/
3 Sourced at www.local.gov.uk/publications/public-private-partnerships-driving-growth-building-resilience
4 Sourced at https://cdn.gihub.org/umbraco/media/4306/gihub_infrastructuremonitor2021.pdf
5 In 2017, OTPP announced the forging of partnerships with construction companies and developers to tap into greenfield project opportunities.
6 Sourced at https://infrastructuredeliverymodels.gihub.org/case-studies/thames-tideway-tunnel/
7 Sourced at www.arup.com/projects/yorkshire-water
8 Sourced at https://infrastructuredeliverymodels.gihub.org/case-studies/heathrow-terminal-5/
9 Sourced at www.slideshare.net/kellygale/international-business-assignment-8
10 See www.hiltonfoundation.org/humanitarian-prize/laureates/aravind-eye-care-system

10 Infrastructure business models

As the infrastructure gap continues to increase, it will become more apparent that existing approaches to infrastructure delivery are not sufficient. We see this happening in the transportation sector where Uber addressed the "last mile" problem and challenged both public transit and regulated taxi service. Uber is just one of many evolving business models based on a "pay as you use" principle and for which a positive customer experience is paramount. These models address some of the current challenges of infrastructure delivery including the high up-front capital costs, maintenance and repair, increased user involvement in on-going improvement, revenue models that are user-friendly, and the ability to capture benefits of additional investment. New business models are also responding to opportunities tied to environmental initiatives such as the impetus to reduce, reuse, and recycle. New business models are not just about accessing private capital. They involve capitalizing on technology and market knowledge. A good example is addressing the transition to net-zero housing by many Western nations over the next two decades that will require significant innovations in how housing is built, maintained, and operated.

New infrastructure models will fall into three general categories (Davies et al., 2010):

1 Shifting responsibility for core infrastructure activities from the public to the private sector. These could involve shifting total responsibility for the design, build, operation, maintenance, and finance to the private sector, with all costs and revenues being in private hands. Governments' involvement might be limited to concession agreements or some form of JV participation. These models could be used in a range of transportation options, energy production and distribution, communications, and water and wastewater systems.
2 Risk-sharing models involving both private and public organizations. This includes new iterations of existing PFI and PPP models. New arrangements could be pursued that provide for the sharing of risks and revenues, rather than just risk transfer such as with the Alliance model.

DOI: 10.1201/9781003396949-13

3   New types of organizational forms and transactional relationships between
    user groups and private partners. These initiatives would be driven by
    efforts to co-develop or co-create new services and added value solutions.
    A basis for these models would be long-term relationships built on trust
    rather than contracts, with an emphasis on the provision of high value-
    added services that meet regulatory and government policy directives and
    pressures.

Nav Canada

Nav Canada is an example of shifting to the private sector the delivery of a
critical public service that was previously the exclusive domain of the public
sector. Nav Canada is a privately run, single-share, not-for-profit corporation
that owns and operates Canada's civil air navigation system in accordance
with the *Civil Air Navigation Services Commercialization Act*[1]. The company
employs approximately 1,900 air traffic control flight service specialists and
700 technologists. It is responsible for the safe, orderly, and expeditious flow
of air traffic in Canadian airspace. It has been in operation since November
1, 1996, when the Government of Canada transferred the entire responsibility
from the federal government to a private firm. As part of the privatization of
what was previously a public service, Nav Canada paid the Government of
Canada CAN$1.5 billion.

Nav Canada is now the world's second-largest air navigation service provider based on traffic volume. It operates independently of any government
funding and is restricted in its funding sources to publicly traded debt and
service charges to aircraft operators. As a non-share capital corporation,
Nav Canada has no shareholders. The company is governed by a 15-member
board of directors representing the four stakeholder groups that founded Nav
Canada—air carriers, general and business aviation interests, the federal government, and the bargaining agents (unions). The four stakeholders elect 10
directors, and these 10 directors then elect four independent directors, with no
ties to the stakeholder groups. These 14 directors then appoint the president
and chief executive officer who becomes the 15th board member.

This structure ensures that the interests of individual stakeholders do
not predominate, and no member group can exert undue influence over the
remainder of the board. To further ensure that the interests of Nav Canada
are served, these board members cannot be active employees or members of
airlines, unions, or government.

Airports themselves are excellent examples of new business opportunities
that utilize the private sector to provide an essential public service. Airports
are a key component of the air transport system. They provide the entire
infrastructure and most processes needed to support passengers and freight
transfers and traffic. Traditionally, airports were seen as dedicated to inter-
modal transport infrastructures, interfacing air, road, railroad, and maritime
modes of transport. Presently, more than being transport infrastructures,

airports are platforms for regional business growth and are now managed by a growing number of international global airport companies.

Airports have become complex enterprises that demand a wide range of business competencies and skills, just as with any other company or industry. Looking at airports as platforms connecting airlines and passengers requires a new business perspective (Brilha and Nobre, 2019). According to the paper by Brilha and Nobre, airports ought to be seen as two-sided platforms able to provide value to both passengers and airlines. More passengers will result in more profit to airlines, and more flights and carriers will better serve passenger needs. In this "multi-sided" market perspective, airport revenues come from airlines (one side) and passengers (other side), while airports assume an important role as a distribution center attracting both sides and getting them connected. Their proposed model is based on the concept of entrepreneurial management, unbundling of services, and capturing value-creation opportunities such as commercial enterprises. Add to the mix the rapid growth of the air cargo and on-site retail services, both significant sources of airport revenues today.

Sharing risks and rewards

Examples of risk and reward sharing are harder to find. Perhaps the best examples are projects that are referred to as social business models in which trust is a factor in the creation of value, as well as a key component of conceptualization and operationalization of a business model (Jablonski and Jablonski, 2020). This is the case with business models in the water supply sector where trust, environmental engineering, industrial ecology, and innovative approaches and solutions play key roles. These are business models that must build both economic and social value and offer a high-trust product. However, this is a topic fraught with political and ideological overtones that are far beyond the scope of this book.

One example of government and the private sector attempting to share risks and rewards, but not always successfully, is the Highway 407 toll road in Ontario, Canada. This is one of the few toll roads in Canada. This 151.4 km long continuous strip is the fourth longest freeway in Ontario's 400 series highway network, all of the other three being free. First launched as a public highway, it is now comprised of a privately leased toll segment (407ETR) and a publicly owned segment, only portions of which are tolled. Tolls are electronically operated by a single firm, and there are no toll booths along the length of the route. Distances are calculated automatically using transponders or automatic number-plate recognition, which are scanned at entrance and exit points. It was planned in the 1950s, construction started in 1987, but in the early 1990s the Ontario Government ran out of funds and pursued a PPP concession agreement on a 99-year lease. The tolled portion (407ETR) is reputed to be among the most lucrative toll roads in the world, not something that the Ontario Government is pleased to hear. The 407ETR is now indirectly owned by a subsidiary of CPP Investments (total 50.01%); Cintra Global S.E., a wholly owned subsidiary of Ferrovial S.A. (43.23%); and SNC-Lavalin (6.76%)[2].

The operator was granted freedom to set tolls based on maintaining a minimum traffic threshold to relieve congestion on alternate routes. Traffic volumes soon exceeded forecasts and the operator quickly realized there was high demand in the system and responded by increasing tolls. What the public see as the "privatization" of Highway 407 has been the source of significant media attention and criticism, especially regarding the annual increases in tolls, license plate denial for non-payment of tolls, and alleged false charges. The Ontario Government took the operator to court to limit fee increases and lost the case.

The saga of 407ETR illustrates the difficulty of reconciling public and private interests over the life span of a concession. The result in this case is a continuous highway of which a portion is privately owned and tolled, portions are publicly owned and tolled, and most recent additions are publicly owned and not tolled. Very recently the government removed tolls from portions they owned. An election was on the horizon. The private portion was built under a PPP arrangement on a revenue-based model, the public portions are also PPPs, but with availability payments. The tolls on the public portion are collected by the one private operator but paid directly to the Province and are set lower than on the private portion. These results can hardly be called an effective public–private interface; rather the example embodies political posturing across various governments in power at the time.

Recent examples of growth strategies

Examples in the third category exemplify opportunities to take advantage of new technologies, pursue new business models, and build platforms for future growth. A new offshore wind business launched by Macquarie's Green Investment Group has teamed up with Ontario Teacher's Pension Plan (OTPP) to develop offshore wind projects (Baba, 2022). Corio Generation is a Macquarie Green Investment Group company, operating on a standalone basis. The wind portfolio consists of 14 fixed-bottom and floating projects in South Korea, Taiwan, Japan, Ireland, and the UK currently under development, with the opportunity to expand the partnership through creation or acquisition of new projects. This investment falls under OTPP's recently established Greenfield Investments and Renewables group. OTPP will acquire a half interest in Corio's portfolio supporting development, construction, and operation.

This is an example of a risk-bearing model in which a private firm delivers clean energy, an essential public service, at no risk to government. Corio, a private company retains full responsibility for the risk of cost and time overruns and failure to achieve desired outcomes. Risk bearing by Corio will depend upon their in-house capabilities to preside over all strategic decisions and act as the systems integrator to coordinate and control the project. This involves a partnership approach to share risk, but not to transfer or avoid risk. Governments in the various jurisdictions in which they build and operate their offshore wind farms maintain their standard regulatory controls but are not involved in any decision-making or risk-taking activities.

A second example takes infrastructure to the new frontier of green hydrogen and the push for decarbonization (Figure 10.1). This technology is based on the generation of hydrogen, a universal, light, and highly reactive fuel, through a chemical process known as electrolysis. This method uses an electrical current to separate the hydrogen from the oxygen in water. If this electricity is obtained from renewable sources, energy will be produced without emitting carbon dioxide into the atmosphere.

As the International Energy Agency (IEA) points out, this method of obtaining green hydrogen would save the 830 million tonnes of CO_2 that are emitted annually when this gas is produced using fossil fuels (Iberdrola). Likewise, replacing all gray hydrogen in the world would require 3,000 TWh/year from new renewables, equivalent to the current demand of Europe. However, there are some questions about the viability of green hydrogen because of its high production cost. Reasonable doubts that will hopefully

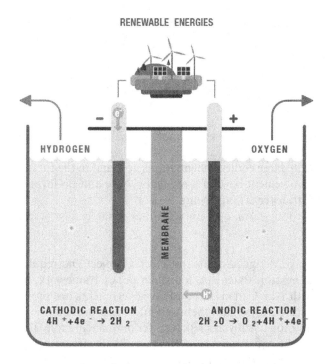

Figure 10.1 Green hydrogen production.

Source: U.S. Department of Energy and Wood Mackenzie.

disappear as the decarbonization of the Earth progresses and the generation of renewable energy becomes cheaper.

A consortium of GIC, the large Singapore sovereign wealth fund, Ontario Teacher's Pension Plan, Alberta Investment Management Corporation, and Manulife, a large Canadian insurance company will fund Haddington Venture's bid to construct the first of a series of green hydrogen platforms in the western U.S. (O'Brien, 2022). The first project will be to develop an underground salt cavern in Delta, Utah, set to become the site of the world's largest green hydrogen platform to date. This investment represents an initial $650 million equity syndication run by Haddington with the opportunity to increase this investment to $1.5 billion. The project plans to be up and running in 2025, generating 220MW of electrolyzer capacity to produce green hydrogen to be stored in salt-dome storage caverns and made available on demand to the Intermountain Power Agency (IPA).

Haddington was formed in 1998 specializing in oil and gas storage facilities and is joint venturing with Mitsubishi Power Americas and Magnum Developments, a portfolio company of Haddington. The U.S. Department of Energy provided a landmark loan guarantee of $504.4 million. The Department of Energy has set aside $8 billion to support 8–10 hydrogen hubs and producers expect the Federal Government to introduce a hydrogen production tax credit to accelerate more projects.

These two examples highlight several common features with investor forays into greenfield projects:

1 Involve the creation of a new asset that can generate long-term revenue.
2 Rely upon strategic partnerships with partners that can share and thereby mitigate some of the risk, and bring skills, knowledge, and specialized expertise to the deal.
3 Represent a platform approach that can support growth and scaling up across a broad geographic area.
4 Capacity to accommodate large equity contributions now and in the future.
5 Limited government involvement beyond a regulatory role and no involve government in any risk transfer arrangements.
6 Result in a standalone long-term operating business that can generate revenue.

New business models must be designed at the outset of a project and require a clear articulation of the mission statement, a distinct policy framework, or even new legislation as with Nav Canada. Too often governments are prone to launch a new infrastructure project and then well into the process look for ways to attract the private sector, usually in a risk transfer or financing role. The design of models suited to the cause must be a "front-end" activity engaging both the public and the private sectors at the outset. Governments must recognize the need for a changing balance of responsibility, the assignment or sharing of risks and revenues, and identify value-capture opportunities that can provide fertile ground for innovation and the launch of creative business models.

Notes

1 Sourced at https://en.wikipedia.org/wiki/Nav_Canada
2 Sourced at https://407etr.com/en/highway/corporate/investors.html#:~:text=407%20
ETR%20%E2%80%93%20Financial%20Information&text=The%20Company%20
is%20owned%20by,SNC%2DLavalin%20(6.76%25).

Part III

The investment universe

11 The global market for infrastructure investments

Determining the size of the global investible infrastructure market requires some guesswork and an understanding of what is, and what is not, included in the estimates. There is much more data on the demand side, particularly with reference to what is needed to close the "infrastructure gap" and this can be broken down by region and sector. Oxford Economics/GI Hub presents two sets of forecasts for global infrastructure investment (GIHub, 2017):

A "**baseline**" forecast to reflect infrastructure investment under the assumption that countries continue to invest in line with current trends, with growth occurring only in response to changes in each country's economic and demographic fundamentals.

An "**investment need**" forecast to demonstrate the investment that would occur if countries were to match the performance of their best performing peers, after controlling for differences in the characteristics of each country. The Oxford analysis suggests that "*if current trends continue, global infrastructure investment will reach $3.8 trillion in 2040, an increase of 67 percent over the 2015 value, in real terms. This reflects the economic growth and demographic shifts that are forecast over the timeframe to 2040*". They also forecast that if "*countries wish to raise their game to match their best performing peers in terms of the resources they dedicate to infrastructure, the forecast value of infrastructure investment need rises to $4.6 trillion in 2040. That is, by 2040 there could be a gap of US$820 billion between what would be spent if current trends continue and what could be spent if all countries matched their best performing peers*".

The report by GIHub/Oxford Economics summarizes infrastructure investment needs and gaps by sector and region, and broken down by current expenditures, needed level of expenditure, and the resulting gap (GIHub, 2017). According to this estimate, sector spending needs are greatest for electricity and roads, which together account for 65 percent of global infrastructure investment for the forecast period under the current trends scenario, or 67 percent under the investment need scenario. The gap between the two scenarios is

DOI: 10.1201/9781003396949-15

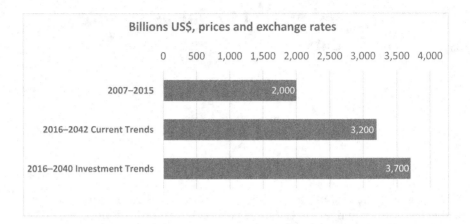

Figure 11.1 Average annual global infrastructure spending requirement, 2016–2040.

Source: Oxford Economics/GIHub. Global Infrastructure Outlook. 2017.

proportionately greatest in the roads and ports sectors, where investment needs are just over 30 percent greater than the estimated spending under current trends. The gap is also relatively large for airports, where the spending requirement is 26 percent greater under the investment need scenario than under current trends.

By region, according to their current trend's scenario, 59 percent of estimated global infrastructure spending needs are in Asia. A further 17 percent relate to the Americas, and 16 percent to Europe. The gap between the two scenarios is greatest for the Americas and Africa, where the forecasts under the investment need scenario are 47 percent and 39 percent greater, respectively, under current trends. More than half of the gap for the Americas is attributable to the U.S. In dollar terms, almost three-quarters of the global infrastructure investment gap between the two scenarios is attributable to Asia and the Americas.

Supply-side data are much more difficult to consolidate and interpret. A significant challenge is to separate public and private expenditures as most data on expenditures refer to public infrastructure projects, almost half of which are estimated to involve some level of private investment, either directly or through PPPs. The Global Infrastructure Outlook estimates that the global infrastructure market totaled $3.1 trillion in 2016 and was expected to grow by an annual rate of 0.6 percent until 2020, to reach $4.2 trillion in 2020 (GlobalData, 2017). Of this amount, developing economies accounted for 62 percent of total infrastructure spending and were forecast to represent 66 percent by 2020. China accounted for 29 percent of the world's infrastructure spending in 2016. China's investment need is more than double that of the U.S.

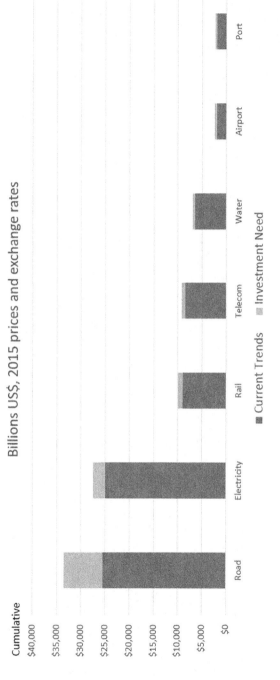

Figure 11.2 Global investment requirements by sector, 2016–2040 cumulative.

Source: Oxford Economics/GiHub. Global Infrastructure Outlook 2021.

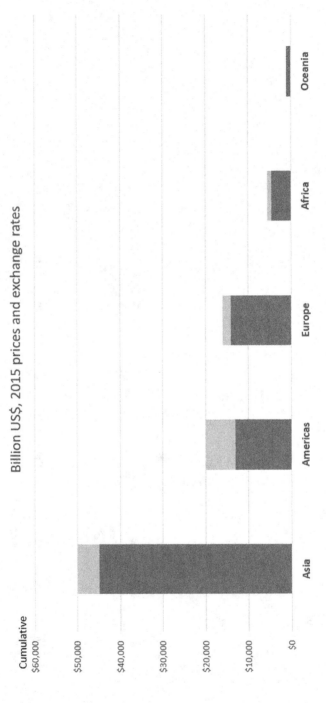

Figure 11.3 Global investment requirements by region, 2016–2040 cumulative.

Source: Oxford Economics/GIHub. Global Infrastructure Outlook 2021.

12 The flow of infrastructure capital

Private capital flows into infrastructure through a network of investment opportunities, with a variety of vehicles that can broadly be defined as either exposure to financial assets, or exposure to real assets. This is a shifting investment landscape driven by many factors:

- The rapid rise in infrastructure investing since being launched around 2000
- The magnitude of need across the globe for new infrastructure
- Refurbishing existing infrastructure
- The search by investors for increased yields
- Diversification beyond what is offered in the equity and bond markets
- The burgeoning market for new infrastructure products to meet ESG targets
- New technologies, particularly in telecommunications

Prior to 2000, private investment in infrastructure was relatively insignificant, beyond corporate spending. Commercial banks played a key role in financing infrastructure projects. As previously mentioned, change began in the mid-1990s following the large-scale privatization and liberalization in the energy, telecommunications, and rail sectors in Europe, attributed in large part to the Thatcher era. Since the turn of the century commercial banks have reduced their involvement in infrastructure except for construction lending, and financing has been replaced by new sources of capital. By March 2017, there were over 165 infrastructure funds in the market, targeting $95 billion in investor capital (Preqin, 2016) (Figure 12.1).

In 2020, the unlisted infrastructure funds raised $100Bn through 101 separate funds (LPs) that closed in 2020 (Preqin, 2021). There was a total of $655Bn in global unlisted infrastructure assets under management (AUM) in June 2020 and an estimated 7.5 percent annualized net return by the unlisted infrastructure funds in the three years to June 2020.

Reflecting the growth in money and the maturation of the fund industry over the past decade is the increasing use of separate infrastructure allocations by the large investors. A recent survey by Probitas Partners indicates that 65 percent of all respondents had separate infrastructure allocations and 69 percent

DOI: 10.1201/9781003396949-16

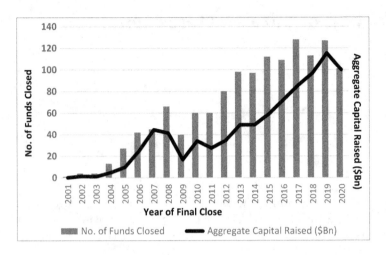

Figure 12.1 Annual unlisted infrastructure fundraising (USD bn).

Source: Preqin.

of experienced investors had such allocation (Probitas, 2019). In 2007, in the first survey in the series, 46 percent of respondents made their infrastructure investments from private equity allocations, while only 26 percent had separate infrastructure allocations.

Figure 12.2 understates total capital going into the infrastructure market since it does not track co-investments and direct investments made by institutional investors. These two activities are much more difficult to accurately track, despite direct investments becoming a more important part of the infrastructure investment market since the Great Financial Crisis. From small beginnings in the mid-2000s, infrastructure funds have become a key diversifier, with investors attracted to the common dual collateral of off-take structures and underlying real assets.

Dry powder

Dry powder is the capital which a company has committed to invest but has not yet allocated. Globally, the uncalled capital commitments of private market funds have been increasing consistently for several decades and reached a record level of $3.1 trillion in February 2021 (Mercer, 2021). This represents a nine-fold increase over the past 20 years. Interestingly, during this same period, assets under management by private market fund managers grew even faster, reaching $7.5 trillion, a more than 11-fold increase.

Investment conduits

- **Open-end funds** focus on core/core-plus assets, with a limited number of funds in the market at any one time. These funds largely rely on a single

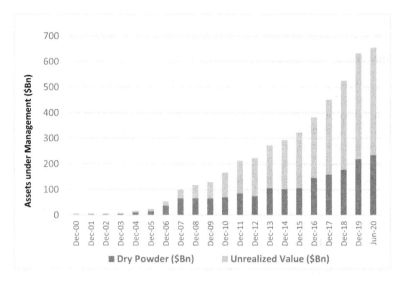

Figure 12.2 Unlisted infrastructure assets under management, 2000–2020.

Source: Preqin.

source of return over the life of the fund. They are possibly more liquid and generally have lower observed volatility than public equities, even in times of distress.

- **Closed-end funds** invest in the full spectrum of core/core-plus/value-added/ opportunistic assets. They will typically have a niche strategy such as sector or region specific. Returns will be based on diversified sources of revenue, including both dividends and capital appreciation. Individual fund terms usually run 10 to 12 years.
- **Fund-of-funds** invest in the full spectrum of core/core-plus/value-added/ opportunistic assets and will also have a niche strategy such as sector or region specific. Most other features are the same as for closed-end funds.

A Probitas survey indicates that interest in fund strategies has shifted beyond core, yields have diminished, and investors are pursuing higher return strategies (Probitas, 2019). While value-added now takes a more prominent role, there is still strong interest in core strategies among direct investors who can avoid management fees. Some fund managers are pursuing what are referred to as "light" funds—investments that are targeting higher returns, more like private equity, but without many of the downside protections that many investors associate with infrastructure. Infrastructure debt fundraising remains low but growing. Renewable energy is the leading industry sector though it is now joined by transportation. North America and Western Europe continue to be the geographies of greatest investor focus, along with global funds whose portfolios usually target OECD countries. Emerging market interest remains relatively weak, with interest in Asia the strongest (Probitas, 2019).

Infrastructure risk and return categories

Infrastructure equity investing is typically divided into four main categories: core, core plus, value-added, and opportunistic (Mercer, 2021). These are commonly accepted classifications used by investors to assist in portfolio construction and dealing with diversification. Classifications are overlayed by additional considerations such as the development stage of a particular asset, whether it is greenfield or brownfield, where it is located, the structure of the deal, and level of involvement in a transaction.

- **Core** is considered the most stable form of infrastructure equity investing. These are assets that are essential to society, are largely de-risked (brownfield assets), and returns are derived from income with limited upside from capital gains. Leverage could be as high as 90 percent. These assets are generally held for the long term (more than seven years), and may be subject to regulation, availability payments, or long-term contracts with highly creditworthy counterparties. Examples are gas, electricity, power generation, mature top-tier airports, toll roads, and PPPs. Net IRR targets may be in the 6–9 percent range, and yield expectations in the 5–7 percent range (Mercer, 2021)[1].
- **Core-plus** is similar to core, but with more variability in cash flows. While income is still the major component of overall return, there is also opportunity for capital appreciation. Leverage may be up to 60 percent. These are primarily brownfield with holding periods usually exceeding six years. Core-plus assets may be much less monopolistic and linked to growth or some other form of asset or contract optimization. Examples include contracted thermal power generation, contracted renewable power generation, contracted oil and gas midstream assets, and toll roads, airports, and seaports with greater GDP sensitivity. Net IRR targets may be in the 9–12 percent range, and yield expectations in the 4–6 percent range.
- **Value-added** are assets that involve expansion, or opportunities for growth or repositioning. Holding periods are typically shorter, in the 5–7-year range. Leverage may be up to 60 percent. A distinguishing feature from core and core-plus is that returns are primarily from capital appreciation rather than ongoing income. They can include greenfield in construction (can be de-risked after construction to fall into core-plus once commissioned), early stage oil and gas midstream, data centers and fiber optic networks, or assets undergoing expansion or repositioning. Net IRR targets may be in the 12–15 percent range, and yield expectations in the 2–3 percent range.
- **Opportunistic** involves assets with a high degree of risk, but also high return potential. Assets may be in development, in emerging markets, in volatile

markets, or in need of significant repositioning. There is a similarity to private equity investments. Holding periods are relatively short, 3–5 years, and returns are almost entirely based on capital appreciation. Examples include assets in emerging markets, or markets undergoing financial distress or transition. Net IRR targets may be in excess of 15 percent, with zero yield expectations, but with implied capital gains expectation in the 15+ percent range.

Investing the proceeds

The proceeds of private investment funds can be used in two ways—invest in existing infrastructure (brownfield) or invest in new infrastructure (greenfield)—and in two forms—equity and debt. Ownership, whether public or private, also influences the capital structure of a transaction with respect to the mix of equity and debt. Unfortunately, the terms funding and financing are often used incorrectly as being interchangeable. This distinction between funding and financing leads to misleading statements by politicians, policy makers, and public interest groups.

Where debt (leverage) is involved, infrastructure is typically financed, as distinguished from funded, through fixed-income products comprising public and private fixed interest instruments such as government-issued or corporate bonds, as well as forms of senior and junior (mezzanine) debt. A distinction is drawn between financing through traditional debt and equity, and financing through project financing. The former is carried on the balance sheets of sponsors or government agencies, and the latter is based on projected cash flows of the facility being financed. The combination of loans and bonds accounts for the most substantial portion of infrastructure finance (Tian et al., 2020). In the equity market, listed infrastructure equity funds, exchange-traded funds, trusts, indices, and unlisted (private) infrastructure funds are the most common sources of infrastructure finance.

Around 80 percent of most private financing of infrastructure projects in the primary markets is provided by financial services institutions such as commercial banks, investment banks, or private equity funds. Many of these transactions utilize forms of project finance where there is a dedicated revenue stream that can be collateralized. In middle- and low-income countries, non-private institutions play a key role as sources of finance. This includes the bilateral and multi-lateral development banks (MDBs), export credit agencies, and governments themselves. However, there is now increasing involvement by the debt markets (GIHub, 2021).

High-income countries account for around three-quarters of all private financing of infrastructure projects with the remaining 25 percent in middle- and low-income countries. The renewable sector dominates in high-income countries, primarily wind and solar, while in low-income countries a significant focus is still on non-renewable energy generation, representing around 25 percent of their total financing.

Several features of infrastructure projects account for the lack of enthusiasm by private investors in greenfield projects, particularly on the equity side. In many high-income countries, new infrastructure consists of a myriad of relatively small non-standardized projects sponsored by cities and regional governments and largely financed from a mixture of capital budgets, government bonds, construction lending from commercial banks, or in the case of the U.S., though tax-exempt municipal bonds. There is very little equity involved and lending rates are relatively low compared to those in private lending. In middle- to low-income countries there is a lack of bankable projects and seldom a pipeline of investment-ready projects. Project preparation is usually led by the MDBs and other international organizations, and NGOs who are involved in financing rather than funding of projects, and seldom involve the private sector at the project preparation stage. The result, from a private investor perspective, is minimal opportunity for private equity, although private debt may be a possibility.

Figure 12.3 summarizes the financing of private investment in infrastructure projects. Private investment in infrastructure by the various private funds can involve significant leverage, mostly with private sector loans. This is the case in high-income countries (78 percent of the total) and with core and core-plus acquisitions. Governments in the emerging markets are largely restricted

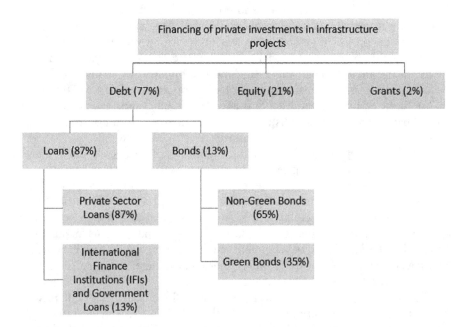

Figure 12.3 Private investment in infrastructure by instrument type, 2010–2020 (% of total value).

Source: Global Infrastructure Hub. Infrastructure Monitor, 2021.

to financing from international finance institutions (IFIs) and government loans, include lending from development banks (multilateral and national), export credit agencies, and the public sector (such as government authorities and state-owned enterprises).

Mapping private investment opportunities

A recent report by the OECD presents a comprehensive, but geographically limited, mapping of infrastructure investments by institutional investors domiciled in OECD and G20 countries, with a view to measuring progress and providing a baseline for future tracking (OECD, 2021). According to this OECD research, institutional investors in the surveyed countries currently hold $1.04 trillion of infrastructure assets. For asset owners, the bulk of their infrastructure investment occurred through unlisted funds and project-level equity or debt, suggesting an illiquidity preference. Investors also exhibited a preference towards assets located within their region of domicile and for cross-border holdings, which are relatively limited. The lion's share is directed towards mature markets. This highlights the critical role of domestic policy frameworks and an investment-grade enabling environment to attract and scale-up institutional investment.

To understand the flow of capital, two Sankey Charts[2], Figure 12.4 and 12.5, depict on the left the origin of investment capital by sources, and on the right the destination of capital by sectors (OECD, 2021). Between the two are the financial instruments through which investment is channeled. Except in the case of direct equity and debt, the ownership relationship between assets and investors is indirect. Depending on the financial instrument used, investors may be economic owners of the assets, legal owners, or both. For example, investors in unlisted funds (LPs) are beneficial owners of the fund's assets, but not legal owners of record. Ownership would likely reside with the GP, a combination of partners, or perhaps some portion remains with the original owner.

These two diagrams provide a snapshot of institutional holdings in infrastructure in 2021 in OECD countries—a total of $3.34 trillion (OECD, 2021). As shown in Figure 12.4, $1.04 trillion is allocated through all instruments (other than listed stocks). Unlisted funds are the dominant conduit of these infrastructure investments, with $380 billion (37 percent) in invested assets. $173 billion is currently held in direct project equity, with $26 billion in direct project debt. Investment through securitized structures including REITS, YieldCos[3], MLPs, and INVITS[4] together represent 43 percent of current institutional investment. As shown separately in Figure 12.5, $2.3 trillion is directly held in listed stocks of companies developing, managing, and/or operating infrastructure assets (listed infrastructure). As discussed above, stock investments do not channel capital to the investee company and therefore cannot direct capital to new investments.

The OECD study indicates the split between primary stage and secondary stage investments, shown in Figure 12.5, which counters commonly held

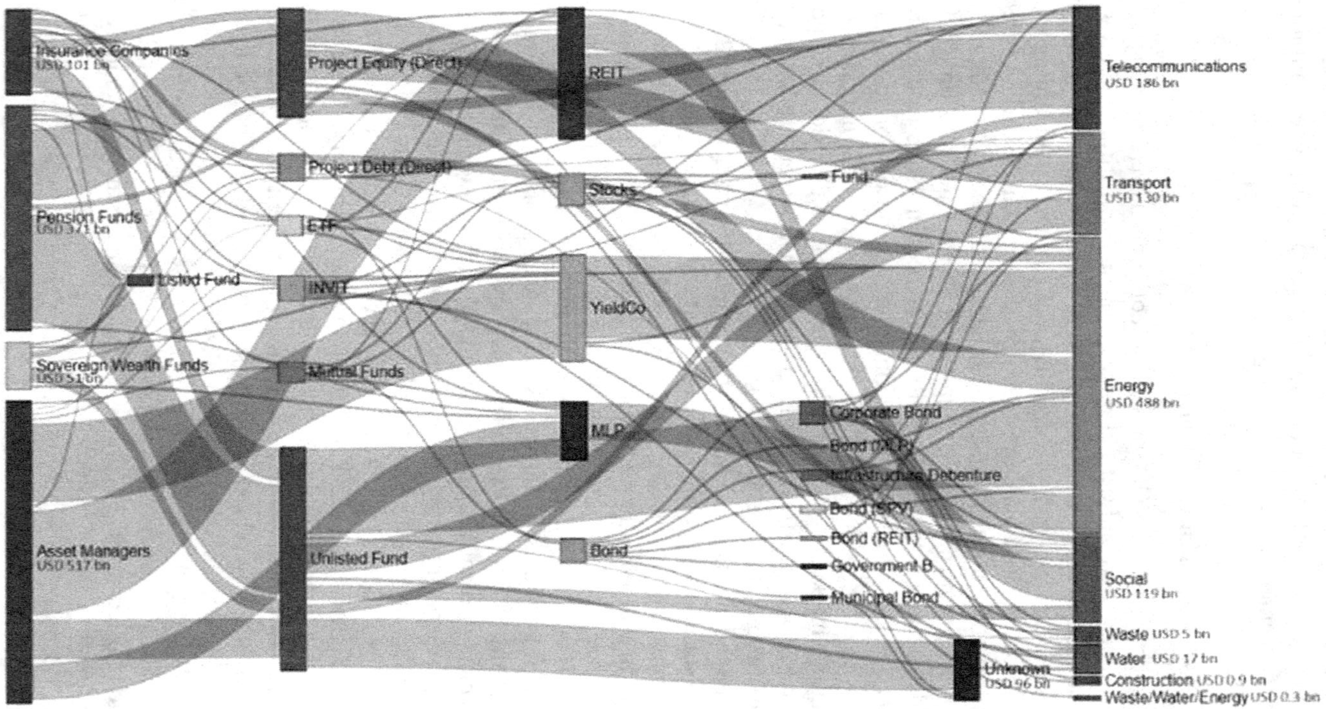

Figure 12.4 Institutional investment in infrastructure (excluding direct investment in stocks)—USD 1.04 trillion. Holdings of institutional investors domiciled in OECD and G20 countries (February 2020).

The figure excludes direct stock holdings (see below). Further, while some nodes appear to have unequal left and right sides, this is just for visual effect, and they are always balanced.

OECD, G... Infrastructure in the Decade for Delivery: Assessing Institutional Investment, 2021

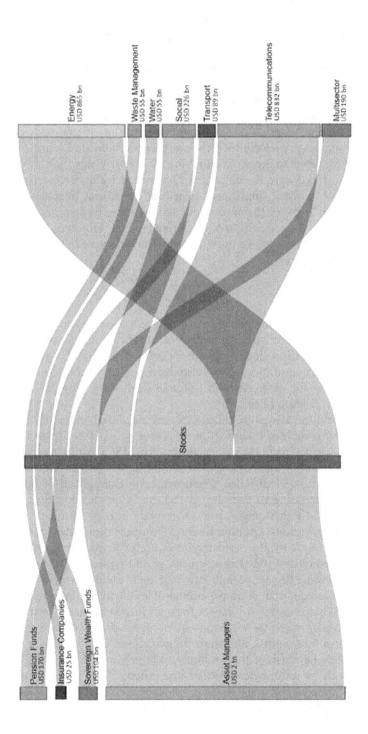

Figure 12.5 Institutional investment in infrastructure-related corporate stocks—USD 2.3 trillion. Holdings of institutional investors domiciled in the OECD and G20 countries) (as of February 2020).

Though some nodes appear to have unequal left and right sides, this is just a visual effect, and they are always balanced.

Source: OECD. Green Infrastructure in the Decade for Delivery: Assessing Institutional Investment. 2021.

beliefs about institutional investors' preferences. Most of the current positions in unlisted funds and direct project-level equity investment are established through secondary stage investment in operational projects. The risk profiles of projects are most elevated during construction phase. However, once projects are operational, project risk is lowered and projects become more palatable to institutional investors. While the preference for operational projects is a longstanding trend, primary stage investment activity by institutional investors has increased in recent years, as in-depth interviews in the OECD study confirm. Construction stage projects with their higher risk-adjusted returns can offer an attractive avenue to investors searching for higher yields.

The most direct exposure to real assets is provided by direct investment at the project level, unlisted funds, and securitized structures like YieldCos, INVITs, and REITs. Most of this investment activity takes place in secondary stage investments, i.e., the acquisition of operating projects. However, the share of debt allocated to primary stage opportunities almost equals debt extended to secondary stage projects.

Measuring performance

Discussing the range of investible infrastructure assets comes back to defining infrastructure and placing it within a context of other options. As previously discussed, there is wide divergence as to what sectors are included or excluded from the term infrastructure and this will also depend upon various economies and those preparing the estimates. An example from data provider MSCI illustrates the point (Figure 12.6).

The MSCI World Infrastructure Index captures the global opportunities of companies that are owners or operators of infrastructure assets (Figure 12.7). Constituents are selected from the equity universe of MSCI World, the parent index, which covers mid- and large-cap securities across the 23 Developed Markets (DM) countries[5]. All index constituents are categorized in one of 13 sub-industries according to the Global Industry Classification Standard (GICS®), which MSCI then aggregates and groups into five infrastructure sectors: Telecommunications, Utilities, Energy, Transportation, and Social.

The standard measure for the performance of an infrastructure investment is the net annualized return, stated as a percentage and typically reported over 1 year, 5 years, or even a 10-year period. However, there are many more factors that must be considered beyond a simple number. The starting point is the inevitable question: What return can I expect from an infrastructure investment? The answer to which might be a barrage of numbers drawn from industry publications, consultant reports, and trade organizations, each with their own reporting purposes. The basis of these numbers is largely unknown and, hence, these numbers can be meaningless (Weber and Alfen, 2010). History offers little guidance. Some of the variables that influence a risk-return calculation include the choice of definitions of infrastructure, the sector, the type of fund, the timeframe (1, 5, or 10 years), the methodology used to derive a particular

Figure 12.6 Ten-year risk-return of different asset classes.

Source: GiHub. Infrastructure Monitor 2021 (MSCI Gross Returns, EDHEC Cintra, Damodaran NYU (US T bonds).

number, and the source of the data. There is also the distinction between net and gross returns which reflects costs and fees involved.

There is a very broad array of infrastructure investment opportunities, investment vehicles, and forms of investment. It is very difficult to draw general conclusions about a particular investment profile without making a precise differentiation between the various possible investment types. The infrastructure investing market is trending toward two camps—a lower risk, lower return market at one end of the spectrum and at the other, a market with investors pushing the boundaries of what infrastructure is and taking more risk in search of increased returns (Campbell, 2022). Also, it is difficult to interpret a particular number in the absence of its contribution to an overall portfolio of investments. This is not to say that that information on performance is not useful in a broader discussion of an overall investment strategy and portfolio construction. It is important to analyze and understand the risk-return profile of a particular infrastructure investment, its correlation to other asset classes, volatility, and diversification attributes.

It is also important to note that institutional investors, and particularly pension funds and insurance firms, are usually required to report annually to their beneficiaries and pensioners. The unlisted funds will use return data in their marketing programs to calculate performance-based fees, guide their investment allocations, and calculate incentive-based payments to employees.

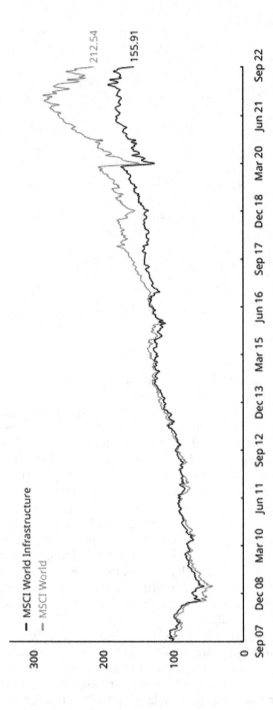

Figure 12.7 Cumulative index performance—gross returns (USD) (Sept 2007–Sept 2022).

Source: MSCI. The MSCI data contained herein are the property of MSCI Inc. and/or its affiliates (collectively, "MSCI"). MSCI and its information providers make no warranties with respect to any such data. The MSCI data contained herein are used under license and may not be further used, distributed, or disseminated without the express written consent of MSCI.

The problem is that there are no reliable benchmarks for infrastructure investing as empirical data are hard to access. Data reside in sources that are confidential. Despite the lack of reliable data, the financial industry tells a story about the benefits of infrastructure investment, regulators increasingly treating infrastructure more favorably than other private assets, and many institutional investors echo these views in their own statements of why they invest in infrastructure (Andonov et al., 2021). The question remains: Is infrastructure delivering the cash flows and returns consistent with the story they tell?

What the research tells us

The research paper by Andonov, Kräussl, and Rauh focuses on the closed-end private funds which represent the lion's share of investor commitments, and the majority of the dollar value of transactions. The authors analyze the risk return characteristics of a group of infrastructure investments, as well as the drivers of their payout policy and performance. In their words, they were testing the hypothesis that "infrastructure investing through closed private funds on average delivers more asset and diversifying cash flows than other alternative asset classes. Instead, we find that infrastructure investment, as institutional investors primarily practice it, has procyclical cash flows generated largely by quick deal exits. Despite the fact that infrastructure covers long-lived tangible assets, the business model of closed funds does not translate any potential differences in the underlying assets into different risk-return properties. We provide three main points of evidence that closed end funds do not deliver on the promised characteristics of infrastructure as an asset class".

The three points in their research findings are summarized below:

1 Based on the risk-return profile of closed infrastructure funds, the average public market equivalent (PME) of infrastructure funds is 0.93, which is lower than the public market equivalent (PME) of buyout, venture capital (VC), and real estate funds[6].
2 Second, the risk of closed infrastructure funds is similar to the risk of other private funds, as their cash flows and returns also primarily reflect quick asset sales rather than long-term stable dividend yields. The closed fund structure in infrastructure provides incentives to exit the best performing assets quickly, similar to the incentives in other private markets.
3 Third, like other private funds, infrastructure funds deliver procyclical cash flows. Net cash flows delivered by infrastructure funds are high when the price–dividend ratio is high and when the yield spread is low. Sensitivity to changes in the market conditions is similar to the sensitivity of other alternative asset classes. Increasing the number of funds reduces idiosyncratic volatility, but there is no additional benefit in terms of cash flow volatility from diversifying across fund types into infrastructure.

The authors point out that, despite weak risk-adjusted performance and failure to match the supposed characteristics of infrastructure assets, closed funds have received increasing commitments over time, particularly from public investors. Also, public institutional investors perform worse than private institutional investors. What explains this underperformance? The authors posit the question: Why then have private infrastructure investments increased dramatically despite this evidence of relatively weak performance?

They point to ESG considerations and a focus on sustainability and impact investing as contributing to increased infrastructure investment overall, and by public institutional investors, but also to the underperformance of public institutional investors. They claim that ESG preferences and regulations explain 25–40 percent of their increased allocation to infrastructure and 30 percent of their underperformance. The research points to two significant implications of this failure of closed funds to match investor expectations. First, the potential underperformance of these funds implies that, over a long horizon, infrastructure runs the risk of not being able to attract sufficient capital in competitive private markets. Second, investors will be less willing to be involved in closing the infrastructure gap.

Analyzing the data

The largest groups of institutional investors are public and private pension funds, sovereign wealth funds, and insurance companies. There are a select group of public pension funds and sovereign funds that invest directly in infrastructure assets and look upon these acquisitions as business propositions, similar to the large equity funds. They undertake their own due diligence and financial analysis of any transaction they are considering, and this is confidential information. All other institutional investors rely primarily on closed infrastructure funds as their investment intermediaries and will source third-party data such as Preqin data sets which are recognized as among the most widely used and comprehensive data available[7]. A unique aspect of the Preqin data set relative to other data sets on private markets is the information on the characteristics of the underlying deals (Andonov et al., 2021). Another source of performance data is EDHE infraMetrics that provides a widely used unlisted infrastructure equity index, valuation and risk analytics, fund strategies, and peer group ratings[8]. MCSI publishes a global quarterly private infrastructure performance index as part of its overall portfolio of indices for developed and developing global investment markets (Figure 12.8)[9].

Measuring the performance of infrastructure investments is a hazardous but necessary undertaking, with levels of uncertainty. But this uncertainty has not slowed the flow of capital into this sector. Quite the opposite. Lack of reliable data will be an ongoing challenge for the industry, and/or for some investors it will make it very difficult to integrate infrastructure into their investment portfolios. This is a situation where market inefficiencies work to the advantage of the large institutional investors who have the capacity to invest directly. It

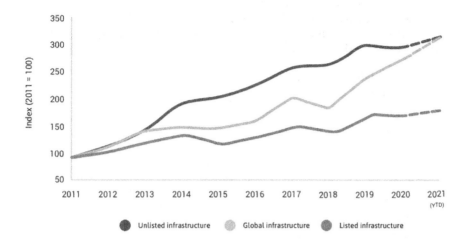

Figure 12.8 Cumulative gross returns performance (index).

Source: Global Infrastructure Hub. Infrastructure Monitor 2021.

also means that the private unlisted funds have wide latitude to interpret data to their advantage. This is not to imply that they intentionally mislead, rather it reflects the inefficiencies in this market.

Notes

1 According to Mercer, net IRR targets are consistent with return expectations set out by active infrastructure investment managers and will therefore vary from Mercer's long-term capital assumptions. Actual returns will differ from expected returns, and estimates are not meant to be a guarantee of performance.
2 See https://sankeymatic.com/build/
3 YieldCos are entities formed to own an operating asset that produces predictable cash flows, preferably through long-term contracts, and to raise funds by issuing shares to investors. They are widely used for solar and wind power projects.
4 INVITS are infrastructure investment trusts used to monetize revenue generating assets and issue units to investor
5 Developed Markets countries include Australia, Austria, Belgium, Canada, Denmark, Finland, France, Germany, Hong Kong, Ireland, Israel, Italy, Japan, Netherlands, New Zealand, Norway, Portugal, Singapore, Spain, Sweden, Switzerland, the UK and the US.
6 In 2015 Preqin introduced a PME metrics benchmark for the performance of a fund, or a group of funds, against an appropriate public market index while accounting for

the timings of the fund cash flows. When PME values are generated for many funds, they can be used as an alternative measure of ranking fund performance, while controlling for broader market behaviour. See https://docs.preqin.com/reports/Preqin-Special-Report-PME-July-2015.pdf

7 www.preqin.com
8 https://edhec.infrastructure.institute/
9 www.MSCI.com

13 Perspectives on the investment industry

Pension funds around the world are adopting a direct investment approach to infrastructure and see this as comparable to investing in business.
(Ron Mock, former CEO of Ontario Teachers' Pension Plan)

Discussions on investing in infrastructure quickly turn to the fund structures involved in sourcing, investing, and managing investment capital, as dealt with in the previous chapter. Unfortunately, not much is known about the management of these funds, other than what they choose to publish, despite them being the foundation upon which a large part of the business of investing in infrastructure is built. Fund managers who oversee these funds fall into three categories: (1) institutional owners of infrastructure assets; (2) managers of unlisted funds, usually general partners (GPs) overseeing limited partnerships (LPs); and (3) large publicly traded equity funds that are owners of infrastructure assets among a diverse portfolio of investments. The large publicly traded funds will operate as GPs, LPs, and other wholly owned or partially owned corporations. Published information on investment funds reveals a relatively short list of large dominant players among the equity and institutional investors on the one hand, and on the other, a long list of smaller investment organizations of which, surprisingly little is known.

All investment managers have as their primary responsibility a commitment to stakeholders, whether these be co-investors, shareholders, pensioners, or insurance policy holders. Success is measured by returns on assets under management (AUM) generated over time. The expectation is that net returns will be materially above what could be achieved through low-cost passive investment. An additional task for some of the large institutional investors is that of administering disbursements such as pensions or benefit claims.

The strength of investment managers, large or small, resides in people, both inhouse talent and the talent imbedded in those who operate their assets. Investment firms rely upon people to get the very best performance from the capital they invest. One can claim that investing in infrastructure is about investing in things and people.

DOI: 10.1201/9781003396949-17

Conduits for channeling investment funds

The large institutional investors in infrastructure can invest through independent funds such as listed and unlisted funds, or as direct owners. Independent investors have discretion in choosing between listed and unlisted funds, what funds to invest in, and how much to invest and over what period of time. They will consider the level of risk they are willing to take, how to diversify their own portfolios, and what returns they hope to achieve. The unlisted funds will market their products to wealthy individuals, pension funds, insurance companies, endowment funds, or investment funds. These funds typically have a life span of 10–25 years if closed-end funds, and unlimited terms for open-ended funds. The GP is compensated on a fee basis similar to that of private equity funds, although a standard fee structure does not exist. As with equity funds, concerns arise over the extent of the costs incurred and fees paid, potential conflicts of interest, and on what basis to evaluate performance. The ultimate success of the GP is based upon their management and technical skills in identifying and evaluating investment opportunities, undertaking thorough due diligence on any investment, and resolving all significant risks affecting the transactions, plus meeting return expectations for a given fund.

Figure 13.1, drawn from the research by Alexander Andonov et al., depicts the investment structures through which 1,861 institutional investors in their database gained exposure to infrastructure assets (Andonov et al., 2021). Investment activity was spread across 5,907 different assets. The study presents a breakdown of the total number of investor-funds managed by GPs—closed, listed, and open-ended—as well as the total number of observations by investor type, plus a transaction count within each type of fund. The study also summarizes the type of institutional investor in each category.

Unlisted closed funds dominate, with 85 percent of the total GP-managed funds included in the survey, followed by 8 and 7 percent, respectively, for the listed and unlisted open-end funds. Pension funds dominate in the category of direct investments. Of the total direct deals observed, 48 percent involved pension funds.

Listed infrastructure investments can incorporate a wide range of investment instruments including shares listed on a stock exchange, shares in listed investment funds, or index certificates (Weber and Alfen, 2010). Listed funds are offered in the retail market as well as marketed to investment organizations. Investments can range from small to very large amounts. The advantage to investors is that they offer liquidity to purchasers and are priced daily so valuation is not an issue. The trade-off is that they are subject to market volatility and performance may correlate closely with other asset classes.

Direct investments and co-investments are done with organizations that are run as businesses that may acquire other businesses. These organizations have overhead costs, risks, and responsibilities that only large institutional investors such as pension funds, sovereign funds, or insurance firms can manage.

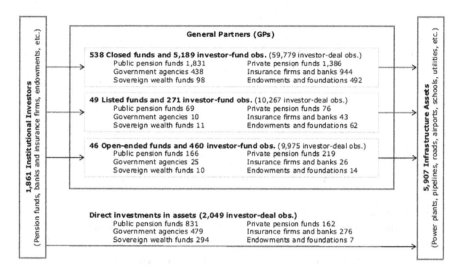

Figure 13.1 Investing in infrastructure.

Source: Institutional Investors and Infrastructure Investing, The Review of Financial Studies. 2021.

They acquire operating businesses across the globe, of which infrastructure will represent only a portion of a total portfolio. These organizations can be involved a multitude of functions from collecting pension funds or insurance premiums, to managing a diverse portfolio of businesses, or paying pensioners or insurance policy holders.

The businesses that direct investors may acquire will still be required to focus on short-term profitability, but institutional owners will take a long-term view of the overall performance they expect to see. They will exhibit patience and determination in investing large sums of money, and not be as swayed by current market conditions as they will be by future growth prospects. Business success will flow from market development, constant oversight, informed governance, continuous improvement, and extending market reach. Entrepreneurship, practiced within defined risk/return parameters, is central to their success.

Institutional investors

No two of the large institutional investors are alike. Each is a highly independent organization with very different organizational structures, what they invest in, their decision-making authority, and reporting relationships. Each fund will be shaped by their mandate, the level of regulatory control they are subject to, their governance structure, and the profile of the pension holders or other stakeholders that they serve. All have organizations that are quite complex and encompass multiple business units. They will have the majority of their assets in global versus domestic markets with field offices located in key overseas markets.

Institutional investors have a large cadre of diverse inhouse professionals and acquire specialized professionals to strengthen their talent base in particular sectors. Their mission is driven by their fiduciary responsibilities to their pensioners or their policy holders. They must ensure that depositor funds, pension receipts, or insurance premiums are invested at rates of return that can provide the benefits expected by their clients. In the case of pension funds, the primary source of funding their obligations will be investment income. For example, Ontario Teachers' Pension Plan (OTPP) funds 80 percent of pension payouts with investment income, the remaining 20 percent coming from employee contributions.

Institutional investors invest capital in three primary buckets: equities, fixed income, and real assets. A portfolio is constructed and managed to achieve an annual growth rate to match or exceed a net return needed to pay pensions based on their actuarial forecast. Portfolios are structured in two dimensions—a strategic allocation to meet performance requirements, and an allocation assigned to capture new investment opportunities. This infers a decision-making process that is top-down in terms of risk management and bottom-up in terms of seeking new investment opportunities. New opportunities are most likely to arise through off-shore offices and benefit from relationships and networks cultivated from having "boots-on-the ground" in their chosen markets.

An investment portfolio that provides the framework for investing in infrastructure will be structured on the basis of three components:

- **Fixed income** that provides security and steady income, hedges against interest rate risks, and stabilizes total returns. Fixed income is expected to have lower volatility and less risk than equities and real assets, and a correspondingly lower yield.
- **Equity investments** that fall into two categories: equity markets and private equity, each with a different risk profile. The combination of equity markets and private equity has the potential to deliver long-term investment growth and value-added performance. Equity investments are usually the largest of the three investment components.
- **Real assets**, the newcomers to investment portfolios, and usually the smallest of the three in terms of assets under management. This category includes real estate and infrastructure, the combination of the two offering a hedge against an obligation to paying inflation-adjusted pensions.

In 2021, the world's 50 largest institutional investors committed $465 billion to infrastructure, nearly $70 billion more than in 2020. Almost half of this amount was committed by the top 10 institutions. In 2022, the top 10 committed $289.6 billion to infrastructure, which represents a total allocation of 5.2 percent of this group. Canadian institutional investors dominate the list

with four of the top 10 positions (42.2 percent of the total allocation to infrastructure), including 1st and 2nd rank (Table 13.1). The average allocation for
the Canadian funds was 14.8 percent, compared to 5.0 percent for the other six
investors. OMERS has the largest allocation, as a percentage of total assets,
at 26.4 percent and they recently announced that they intend to increase this
allocation by 20–25 percent by 2027 (Bentley, 2022).

Ranking second on the list, Caisse de dépôt et placement du Québec
(CDPQ), offers a glimpse into what their business model entails. As depicted
in their Annual Report 2021, Figure 13.2 summarizes the cycle of collection
and payout that defines the scope of their activities.

Table 13.1 The 10 Large Institutional Investors in Infrastructure, 2022

Rank	Investor	Country	Infrastructure Asset (US$000)	Total Assets (US$000)	Allocation to Infrastructure
1	CCP Investments	Canada	38,828,300	411,579,980	9.4%
2	CDPQ	Canada	37,974,077	304,413,872	12.5%
3	Allianz	Germany	37,822,160	1,290,256,560	2.9%
4	ADIA	UAE	37,305,000	829,000,000	4.5%
5	AustralianSuper	Australia	25,903,492	187,564,498	13.8%
6	National Pension Service	South Korea	25,393,277	797,718,038	3.2%
7	OMERS	Canada	25,037,024	94,671,247	26.4%
8	APG	Netherlands	22,705,752	720,206,400	3.2%
9	Ontario Teachers'	Canada	20,420,823	189,029,531	10.8%
10	AXA	France	18,211,789	716,895,262	2.5%
	TOTAL for TOP 10		**289,601,694**	**5,541,338**	**5.2%**

Source: IPE September/October 2022 (magazine).

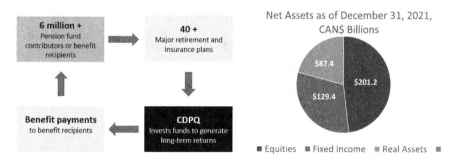

Figure 13.2 CDPQ's business cycle—parameters for investment decisions.

Source: CDPQ Constructive Capital 2021 Annual Report.

The net asset value of CDPQ's total investments in 2021 was CAN$365.5 billon, achieving a 7.7 percent return with an annualized return of 7.8 percent over 5 years. In 2021, returns increased to 13.5 percent in the 1 year and 8.9 percent over 5 years[1]. This number was likely to fall in 2022 in response to the turmoil affecting global markets.

There are three primary challenges for the large institutional investors. For the pension funds it is the demographics of pensioners within their plan; for insurance companies, the changing profile of insured risks and the actuarial profile of the insured; and for the sovereign funds, their return thresholds. The impact of changing demographics on pension funds is illustrated in Figure 13.3 which depicts Ontario Teachers' Pension Plan (OTPP)[2]. OTPP administers pension payments of over $6 billion annually. It had 184,000 active Ontario public school teachers in 2018 and the pension plan served 145,000 pensioners. Average age was 72 years old, with 133 pensioners 100 years or older. A large proportion of pensioners are female, a cohort with increasing longevity. In 1990, there were four teachers for every pensioner with an average pension life of 25 years. By 2018, the ratio was down to 1.3 active teachers for every pensioner, and an average pension life of 32 years.

OTPP sees their primary challenges on the investment side as two-fold: strategic capacity—structuring a portfolio of investments to achieve long-term growth rates and targeted net returns; and implementation capacity—the ability to source, execute, and manage investments in the portfolio. An enterprise risk

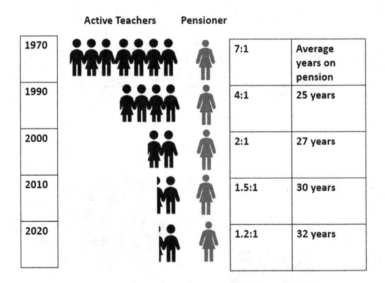

Figure 13.3 Pensioner profile, active teachers to pensioners, OTPP, 1970–2020.

Source: OTPP Annual Reports.

management culture permeates the large institutional investors, spread across five categories: governance risk (brand and reputation are essential); strategic risk (amount and type of risk the organization is willing to accept); investment risk (that must be identified and priced); operational risk (having the right people and processes in place); and reputational risk (expectation of stakeholders and the market).

These challenges can only be met with long-term commitments that can ensure sufficient investment proceeds to meet future funding obligations. Institutional investors must look beyond annual changes in asset values, which are largely unpredictable, and focus on the long-run average profitability of their total portfolio of assets. This highlights a particular feature of institutional investors in managing a portfolio on investments, a feature that is not likely to encumber the listed and unlisted funds. New acquisitions should be accretive and result in gradual or incremental growth in the total value of the portfolio. This accretive value should exceed the cost of acquiring the asset. The reverse is true for dispositions.

Despite an attractive offer for one of its assets, an institutional investor must measure the potential impact of a disposition on the long-run average profitability of the portfolio. A disposition, no matter how attractive, will leave a "hole" in a portfolio that must be filled, now or sometime in the future. This implies an opportunity cost for any disposition that must be priced against the benefits of the disposition. This does not preclude dispositions based on a need to recalibrate diversification targets, reduce certain types of risk, meet regulatory requirements, adapt to new investment objectives or a revised investment strategy, or changing market conditions. A common strategy for both acquisitions and dispositions is entering into new partnership or co-investment arrangements depending on the objectives to be met.

Listed funds (infrastructure managers)

The influence of the listed funds continues to grow as many of the large players in this space are transitioning into private equity firms and raising increasingly large sums of money (Table 13.2). The assets under management of the 100 largest infrastructure investment managers rose from $1.38 Trn to $1.76 Trn during 2021 (IPE Real Assets, 2022). This annual growth of $384 Bn took place while the world was still dealing with COVID and which the transportation sector was particularly negatively affected by. The top 10 in the rankings manage about $817 million between them, representing nearly half of the aggregate AUM of the top 100. The top two, Macquarie and Brookfield Asset Management, manage $415 Bn between them, almost a third (30 percent) of the top 10 total. Macquarie dominates in Europe and Asia-Pacific, and Brookfield dominates the Americas.

The mission of the listed funds is three-fold; to attract capital to grow assets under management; to achieve attractive risk-adjusted returns on invested

Table 13.2 The 10 Largest Infrastructure Managers, 2022

Rank	Company	Infrastructure AUM $Bn	Total AUM $Bn	As Of
1	Macquarie Infrastructure & Real Assets	230,217	251,340	31/12/21
2	Brookfield Asset Management	157,896	459,505	31/12/21
3	Global Infrastructure Partners	76,647	76,647	31/12/21
4	IFM Investors	60,489	115,064	31/03/21
5	Allianz Global Investors	47,244	637,000	31/12/21
6	Stonepeak Partners	42,823	42,823	31/12/21
7	Digital Bridge	42,200	42,000	31/03/22
8	I Square Capital	37,095	37,095	31/12/21
9	KKR	35,318	430,080	31/12/21
10	BlackRock Real Assets	35,054	8,838,465	31/12/21
	TOTAL for TOP 10	**764,893**	**10,930,019**	

Source: IPE Real Assets. July/August 2022.

capital; and to increase shareholder value. Growth reduces overhead costs through economies of scale, increased fees, and access to additional opportunities to invest their own capital. Attractive risk-adjusted returns means exceeding the market as measured by various stock indices. Shareholder value is represented by both the capitalized value of the listed company, and dividend payouts to shareholders. The listed fund managers may be structured as GPs, utilizing various LP funds to invest in both equity and debt. Income from real asset investments is derived from buying and selling assets, management fees, arbitrage transactions, and return on investment of their own capital. A closer look at the two largest funds on the list reveals two very different business models.

Macquarie Infrastructure and Real Assets (MIRA)

Macquarie Infrastructure and Real Assets (MIRA) is the world's largest infrastructure investment manager. The origins if MIRA date back to 1969 and the founding of Hill Samuel Australia Limited, a subsidiary of Hill Samuel & Company Limited in the UK. As a result of financial deregulation and the floating of the Australian dollar, Hill Samuel emerged as Macquarie Bank Limited in 1985 as an Australian financial institution. In 1994, Macquarie launched its infrastructure investment business with a single acquisition and subsequently recognized the opportunity to raise funds from third-party investors that were competitive in terms of cost of capital compared to what the investment banks were offering from a structured finance perspective. Macquarie Infrastructure & Real Assets Inc. was founded in 2007 as one of five principal operating groups within Macquarie Group Limited and falls under Macquarie Asset Management (Figure 13.4). Macquarie Infrastructure

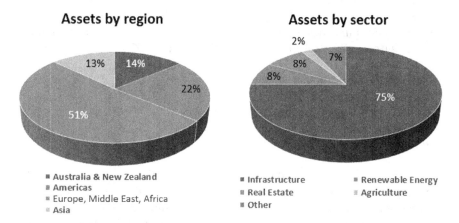

Assets by region

- Australia & New Zealand
- Americas
- Europe, Middle East, Africa
- Asia

Assets by sector

- Infrastructure
- Real Estate
- Other
- Renewable Energy
- Agriculture

Figure 13.4 Macquarie Infrastructure and Real Assets (MIRA), 2021.

Source: Macquarie Asset Management. Putting the world's long-term savings to work. 2021.

and Real Assets (MIRA) represents infrastructure, real estate, agriculture, and energy with $230.2 billion (2022) of assets (92 percent of total AUM).

Brookfield Asset Management (BAM)

Brookfield Asset Management's roots go back more than 100 years to an investment in transportation and electricity. Today, it is the second largest global investor in real assets with investments in real estate, renewable power, infrastructure, and private equity assets[3]. Brookfield, rising from the ashes of Brascan in the early 1990s, is just over 20 years old and is an indication of just how young this industry is, its rapid growth, and the dynamism that surrounds its activities. Brookfield's predecessor can be traced back to 1899, to a firm operating in construction and the management of electricity and transport infrastructure in Brazil. It was incorporated in Canada in 1912 as Brazilian Traction, Light and Power Co. to develop hydroelectric projects in Brazil. In 1969 the name was changed to Brascan Limited.

The sprawling empire that was subsequently assembled imploded during the recession of the early 1990s and it was forced to sell or liquidate most of its holdings. In 2000 the name was changed to Brascan Corp. and under new leadership Brascan sought to remake itself and shed the remaining vestiges of its conglomerate past. The new organization chose to concentrate on real estate, power, infrastructure, and fund management. In 2005, the name was again changed to Brookfield Asset Management (BAM).

Brookfield Asset Management, the umbrella organization with operations in over 30 countries and approximately 150,000 operating employees, has over $495.5 in assets under management (2022) of which 34 percent are in infrastructure ($157,895). It is the significant owner of five listed limited partnerships through which it operates[4]. In 2019, Brookfield acquired a majority interest in Oaktree, and Oaktree continues to operate as a standalone business with assets under management of $159 Bn.

- Brookfield Infrastructure Partners L.P.
- Brookfield Renewable Partners L.P.
- Brookfield Business Partners L.P.
- Brookfield Reinsurance Partners L.P.
- Brookfield Property Partners L.P.

Until spun off in January 2008 as Brookfield Infrastructure Partners L.P., Brookfield Infrastructure was an operating unit of BAM which still retains a 30 percent ownership and acts as the partnership's general manager and financier. Brookfield Infrastructure Partners owns and operates a global network of infrastructure companies in utilities, transportation, energy and communications, transmission and telecommunication lines, toll roads, ports, and pipelines.

Through Brookfield Private Funds, Brookfield typically invests its own capital alongside its investors. Over 30 percent of the capital under management across Brookfield's private funds is their own capital. They are typically the largest investor in any of their transactions. Through BAM, they have the capacity to serve as their own source of funds as well as risk insurers[5]. BAM, the corporation, is currently converting into two publicly traded companies, one being a pure-play asset manager, and the other focused on deploying capital across operating businesses and compounding that capital over the long term[6].

Unlisted funds

From small beginnings in the early 2000s, infrastructure funds have become a key diversifier in portfolio structuring, with most of the growth in the unlisted funds. Unfortunately, as private funds there is a lack of information on this sector and not a lot is known about them except what is published on web sites and in trade journals. There are publications such as Preqin and IJGlobal that track a growing universe of investors and fund managers active in unlisted funds.

According to Preqin, under 100 fund managers have raised the lion's share of capital commitments in recent years (Figure 13.5). Managers that raised four or more funds secured 75 percent of the aggregate capital raised in 2019. Dominance by a select few seems to characterize all three conduits for sourcing investment funds. Experience seems to foster confidence among those willing to entrust their money to managers. It is difficult to make an argument that it is performance that counts in light of the paucity of data.

Unlisted funds are managed by GPs, each a unique entity competing in an increasingly crowded field. These fund managers focus on sourcing, evaluating,

Figure 13.5 Number of active infrastructure fund managers by vintage of first fund closed.

and managing private market investments globally. They are not as actively involved in operating aspects of their assets under management as are institutional investors and some of the listed funds. Income is derived from fees and a carried interest, and therefore they rely on their ability to raise and then invest capital within the structure of mainly closed-ended funds. Preqin reported a mean management fee across the unlisted funds of 1.29 percent on a fund basis in 2020, and a carried interest of 20.00 percent. This fee structure may change with impending tax reforms, particularly in the UK with respect to capital gains, pushing the need for alternative fee structures in anticipation of less favorable tax treatment of investment value uplifts.

Public pension plans in Canada: The "Canadian Model"

To put Canada in an international context, a study by the Thinking Ahead Institute (2021), covering 22 global pension markets, ranked Canada number 4, with total estimated assets of $3,080 billion (ratio of 192.5% of assets/ GDP). For a relatively small country, around 36 million people, there must be something that Canadian pension funds are doing that others are not doing.

The study estimates that of total assets for the 22 markets in the survey totaling $52,522 billion, 92 percent are in the seven markets identified in Table 13.3 (Thinking Ahead Institute, 2021). While the U.S. dominates the list with estimated assets of $32,567 billion (ratio of 158.5% of assets/GDP), one would expect that with this large base of capital the U.S. would be a dominant player in pension fund business. It is not. However, the U.S. is still the preferred investment market for investors across the globe. Within the seven top markets, direct contribution (DC) pension plans represents almost 53 percent of pension assets, and are now dominant in Australia, the US, Japan, and

Table 13.3 Total Estimated Pension Fund Assets, Seven Largest Markets, 2020

Market	Total Estimated Pension Funds' Assets ($ billions)
US	$32,567
Japan	$3,613
UK	$3.564
Canada	$3,080
Australia	$2,333
Netherlands	$1,900
Switzerland	$1,163

Source: Thinking Ahead Institute, 2021.

Canada. The trend to DC is growing with a 20-year growth rate of 7.8 percent per annum relative to 4.1 percent for Defined Benefit (DB).

Public pension plans in Canada have gained international recognition for their size, dominance in international rankings, performance, and commitment to alternative assets including infrastructure. Australia and Canada have regulatory frameworks that allow them to invest with enormous flexibility, and apparent success (Alonso, 2015). Nearly all of the pension funds in Canada are created by federal and provincial legislation that sets mandates and assigns oversight of the activities of a pension fund to an independent board of directors. All funds report to a pension regulator. Among the "Big Eight" funds listed in Table 13.4, there are three instances where the same organization is responsible for both assets and liabilities. Pension boards are accountable to either federal and provincial ministries either directly, or through government departments or agencies. All eight are subject to governance structures which entrusts their boards and their respective investment committees with the crucial responsibility for establishing and controlling the fund's appetite for risk, investment policy, and their respective risk-management frameworks (Bédard-Pagé et al., 2016).

The Big Eight stand apart from most other Canadian and many foreign pension funds by size, how they operate, and their investment strategies. This has given rise to an asset-management approach that some refer to as the "Canadian Model" (Bédard-Pagé et al., 2016). The characteristics of the model are summarized below:

• Internally managed, made possible by economies of scale
• Investment strategies designed to capture the liquidity premiums offered by less-liquid alternative assets
• Diversification across a broader set of asset classes, investment styles, and geography
• Use of leverage and derivates designed to improve returns and mitigate risks
• Reliance on in-house risk-management functions
• Compensation competitive with private sector to attract and retain talent

Table 13.4 Canada's Big Eight Investment Assets Under Management by Rank, 2021

Fund	Net Assets $ Bn 2021	Net Return 2021	5-Year Average Net Return	Percent in Infrastructure
Canada Pension Plan Investment Board (CPPIB)	497.2	20.4%	11.2%	8.3%
Caisse de dépôt et placement du Québec (CDPQ)	419.8	13.5%	8.9%	11.0%
OMERS (Ontario Municipal Employees Retirement System)	120.7	15.7%	7.5%	20.0%
Ontario Teacher's Pension Plan (OTPP)	241.6	11.1%	8.4%	11%
Health Care of Ontario Pension Plan (HOOPP)	114.4	11.28%	11.6% (10yr)	2.0%
British Columbia Investment Management Corporation (BCI)	199.6	7.4%	8.3%	9.5%
Public Sector Pension Investment Board (PSPIB)	204.5	18.4%	9.3%	9.0%
Alberta Investment Management (AIMco) Board	168.3	19.0%	7.4% (4 year)	6.0%
TOTAL Assets Under Management (Big Eight)	$1,966.1 Bn			

Source: Various annual reports.

Although pension funds target something in the range of a net 4 percent annually over the long term to meet their obligations, real returns have historically far exceeded this amount.

Public pension plans in the U.S.

The U.S. is the world's largest economy, with the world's most sophisticated capital markets, most attractive and diverse investment opportunities, and home to many of the world's most sophisticated and successful investors. It is also the largest pension market in terms of asset under management by pension fund, almost 10-fold larger than Japan, ranked in second place. Until recently, almost all public pension plans in the U.S. were in a negative cash flow position, meaning that benefit payments exceed contributions (Lipshitz and Walter, 2019).

With a few exceptions, U.S. public pension plans have consistently underperformed. Public pension plans in the U.S. are predominately defined benefit (DB) compared to most private plans that are defined contribution (DC), so shortfalls represent significant financial liabilities on the books of governments. The U.S. public pension fund system surveyed for a 2017 study, identified 297 distinct state plans and 5,232 local plans with aggregate assets of $4.33 trillion. These funds are held in trust to fund future pensions benefits accrued to the end of 2017 for 20.6 million employees of state, county, and city governments. Pension obligations at the end of 2017 were estimated at $5.96 trillion, indicating a funding shortfall of $1.63 trillion. This is a funding ratio of 72.6 percent (Lipshitz and Walter, 2019). The state retirement systems in the U.S. finished the 2021 fiscal year in their best condition since the Great Recession of 2007–09 (PEW, 2021). Given the current economic climate this may no longer be true.

The aggregate funding ratio of the 100 largest U.S. public pension plans reached 85 percent as of June 30, 2021, according to an estimate from the latest Milliman Public Pension Funding Study (Kozlowski, 2021). The estimate represents a significant improvement from the estimated funding ratio of 70.7 percent a year earlier, thanks primarily to excellent market returns. Estimated assets as of June 30, 2021, totaled $4.82 trillion, up from $3.9 trillion the year before, while estimated liabilities totaled $5.67 trillion, compared to $5.5 trillion the year before. The estimated unfunded liability of $850 billion is the first time Milliman's estimate has fallen under $1 trillion since 2012, according to the study. The Mercer CFA Institute Global Pension Index 2021 (Mercer, 2021) gave the U.S. a rating of C+ with an Index value of 60–65[7].

Pension plans in the U.S. are many, relatively small, usually not well staffed, poorly compensated, loosely governed and in many cases by political appointees, and heavily reliant on external advisors and consultants. Other than CalPERS and CalSTRS, few have the capability to take on direct ownership. There is also a lack of transparency in reporting, and fee disclosure by the pension funds is surprisingly opaque, as are the internal costs of managing their portfolios (Lipshitz and Walter, 2021). U.S. pension plans were late in turning to alternative assets and are under-allocated in infrastructure compared with their Canadian or Australian peers. According to Preqin, total investments allocated to infrastructure by U.S. pension funds that disclose their allocation was $68 billion, which represented 1.1 percent of their total AUM. In comparison, Australian superannuation funds allocate 7 percent. Canadian pension funds allocate the highest percentage to infrastructure at 8.4 percent of AUM (Wahba, 2021).

Why are U.S. pension funds so far behind in sourcing infrastructure investments? While most U.S. pension funds now want to invest in infrastructure, there are too few projects to attract them. Infrastructure is crumbling, yet there are limited opportunities for investing, including a very small number of PPPs. The challenge in the U.S. for new infrastructure projects is that they are managed by a complex set of federal and state entities that exhibit monopolistic tendencies at a local level. For example, 95 percent of public highways and bridges are owned by local and states governments. There are 50 states and more than 80,000 local authorities managing these assets (Wahba, 2021). Pension funds, from a practical perspective, cannot approach each of these entities to identify attractive projects.

Notes

1 Sourced at www.cdpq.com/en/news/pressreleases/2021-results
2 OTPP presentation to the Sustainable Infrastructure Fellowship Program, 2020.
3 https://bam.brookfield.com/
4 Brookfield Annual Report, 2018.
5 Brookfield Reinsurance Partners ("BAM Re") is a reinsurance business focused on providing capital-based solutions to insurance companies and their stakeholders, and provides investors greater flexibility to invest in Brookfield.
6 Information taken from the Brookfield Notice of Special Meeting of Shareholders scheduled for Wednesday, November 9, 2022, p. 6.
7 C+ is a system that has some good features, but also has major risks and/or shortcomings that should be addressed. Without these improvements, its efficacy and/or long-term sustainability can be questioned.

14 Public finance challenges

Options for private investing in infrastructure are many, whereas options for investing public funds in infrastructure are few. Public investments in infrastructure favor wealthy nations who are likely to have mature capital markets, sound credit ratings, revenue-generating abilities including fees and taxes, and room to increase debt. Emerging economies have few options other than local banks and forms of concessional financing issued by development finance institutions (DFIs) and non-governmental finance organizations. Compared to commercial banks, the DFIs accept a higher risk in return for addressing beneficial social and/or environmental challenges. What all governments share across the full spectrum of infrastructure needs is the inability to successfully implement a scale of infrastructure production that can begin to close the infrastructure gap. The bottleneck cannot be attributed to the lack of private capital, rather to the challenges that governments face in raising capital on their own or matching the supply of long-term private capital with investable public projects.

Part of the bottleneck is that, in many countries and particularly at the city level, established spending practices and prevailing fiscal philosophies do not often favor using long-term debt to fund long-term community needs. Pay-as-you-go, or front-end approaches are common, combining development levies, accumulated reserves, reserve funds, and federal or provincial capital grants (Fenn, 2022).

The bottleneck has no quick fixes. It must also be acknowledged that, while additional funding for government infrastructure projects may come from private investors, this will not address many infrastructure projects classified as social infrastructure that have no revenue stream or offer the limited appeal of "bond-like" availability payments. Social infrastructure will continue to be paid for largely by taxpayer dollars, including facilities such as military establishments, schools, hospitals, court houses, jails, and highways. Much of this infrastructure will be procured through conventional means, although some countries may use public–private partnerships or design/build/finance arrangements. Emerging economies will continue to rely almost exclusively on PPPs to attract both concessional financing and institutional investors, plus

DOI: 10.1201/9781003396949-18

secure the necessary expertise. For these economies the biggest impediment is the absence of public funds. However, their challenges go beyond financing and include the lack of a pipeline of structured projects that are bankable from an investment perspective.

Public finance challenges often carry a burden of political ideology, manifested in opposition by some governments and citizen groups to any form of privatization or involvement of the private sector. Surprisingly, a lead example is the U.S., where PPPs have failed to gain traction and citizens are highly suspicious of the motive of private partners in the delivery and operation of even the most basic infrastructure such as roads and bridges. On the investment side, private investors have difficulty pricing political risk in infrastructure transactions, given the significant role that state legislatures play as final arbiters on whether a project will proceed to closure, or not. The result is serious deterioration of existing assets and an inconsistent record of closing new deals.

Project preparation and design

The lack of a bankable and an investment-ready pipeline of new infrastructure projects is often considered one of the major bottlenecks in attracting private capital to public infrastructure. This is particularly true in the emerging economies. Enabling an investment-ready pipeline has consistently featured as a top priority of G20 presidencies (GIHub, 2021). Bankability is predominantly determined at the project preparation stage (PPS), involving a complex series of stages that span from conceptualization and feasibility analysis, to deal structuring and transaction support (GIHub, 2019). It is seldom recognized that project preparation is a costly and time-consuming activity, requiring a number of skill sets, and spread over a period that can range from 3 to 8 years, the average being 6 years. This is a cost that has substantially increased over the years, driven by the complexity of new projects and expanding stakeholder interests. Considerations at the project preparation stage can involve new regulations, environmental factors, social issues, governance requirements, and the assessment of new technologies. GIHub estimates that project preparation costs can be as high as 10 percent of total project costs and range from 3 to 5 percent in developed countries and between 5 and 10 percent in emerging economies. These are not easy monies to find and lack of funds imposes severe limitations on creating a pipeline of bankable projects in both developed and developing nations.

A good example of providing key resources required for project preparation is that of the Financiera Desarrollo National (FDN) in Columbia, South America, one of four national development banks[1]. The FDN deals exclusively with structuring and financing infrastructure projects. The bank offers expertise in four areas: structuring and management of infrastructure projects; identifying financing gaps and recommending how to fill them; structuring

debt security issuances in the local capital markets; and addressing risk mitigation strategies and products. FDN was launched in 2011 from what was the
National Energy Finance (FEN), and began its work in 2013, within the framework of the actions taken by the national government to strengthen the institutional capacity and achieve a true strengthening of the national infrastructure.

FDN has as shareholders the national government through the Bicentennial
Group, the International Finance Corporation (IFC), the Sumitomo Mitsui
Banking Corporation, and the Development Bank of Latin America (CAF). It
is an entity linked to the Ministry of Finance and Public Credit[2].

The proverbial question

Public finance challenges raise the proverbial question for any government,
"Are governments, banks, or capital markets best placed to finance infrastructure?" (Ehlers, 2014). This question is an oversimplification and avoids the
complexity of infrastructure projects as compared to many other public capital
projects. The choices available to the private sector to meet their infrastructure
needs far exceed those available to the public sector, as depicted in Figures 14.1
and 14.2.

Figure 14.1 provides an overview of the different alternatives available to
private investors, divided into equity and debt. Equity and debt can be listed
and traded on an exchange (public), or unlisted and traded over the counter
(Croce and Gatti, 2014). In the case of listed equity and market-traded debt,
reference is made to a traditional investment in listed infrastructure. This is the
area where mutual funds and exchange-traded funds (ETFs) have developed
products to be included in the portfolios of retail investors, high net worth

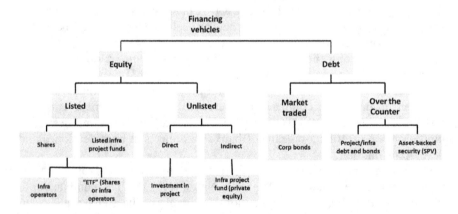

Figure 14.1 Different channels of infrastructure investments available to the private sector.

Source: Raffaele Della Croce and Stefano Gatti. 2014.

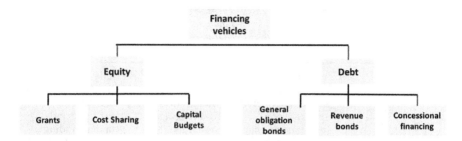

Figure 14.2 Different channels of infrastructure investments available to the public sector.

Source: Raffaele Della Croce and Stefano Gatti. 2014.

individuals, and institutional investors. Unlisted equity or over-the-counter (OTC) debt does not offer the benefit of an active liquid secondary market. OTCs are typical "buy and hold" asset classes, suited to long-term investors with a clear preference for stable, rather than exceptionally high returns.

Figure 14.2 depicts the limited alternatives available to the public sector. These two diagrams highlight the difficulties in matching investment demand to the potential supply of funds throughout the life of a project—it is not simply a matter of financing construction costs. Phases of a typical infrastructure project, all of which can incur costs and require different types of financing, are summarized in Table 14.1.

This matrix depicts the complexity involved in financing infrastructure and the consequences when governments do not have sufficient funds to properly address the requirements of the planning and design phase. This is the advantage of a PPP, whereby a public agency can transfer some of the inherent risks to the private sector, but it is only an advantage if the public sector sponsors have the ability to undertake the necessary due diligence in the project preparation stage and transferred the right risks at the right costs and put in place the appropriate contracts and incentive mechanisms to ensure performance.

Financing of the planning and design stage of public infrastructure will be a cost to be borne by a government sponsor, usually though a grant, a capital budget allocation, a bank loan, or in some cases bond financing with a government guarantee. Bond financing is the least attractive for construction. Commercial bank loans usually supply the largest share of financing in this initial stage. However, such loans come at a relatively high cost and must be retired at the end of the construction period, usually upon reaching substantial completion. This does not preclude an early search by a sponsoring agency for potential equity investors, and perhaps some precommitments on both equity and debt.

It is at the construction phase that equity and debt financing are enabled. This stage also carries some of the most significant risks for which equity holders

Table 14.1 Phases of Infrastructure Projects and Their Characteristics

Phase	Economic justification and contractual issues	Financial characteristics	Potential investors
Planning and design	Contracts with professional service providers spread over a long time period where changes in terms and conditions may require contract renegotiations, additions, and deletions. Rating from rating agencies to secure investor interest, credit insurance, or government guarantees	Search for equity investors who, in turn, need to secure commitments from debt investors (banks)—must parallel planning and design activities over long time periods and involving pre-commitments by debt investors that come at a high cost	Equity investors require a high level of expertise and understanding of key risk factors. Can be construction companies or in rare cases infrastructure funds or direct investments by institutional investors. Debt investors are mainly banks through syndicated loans. Bond financing is rare at this stage
Operations and maintenance	Ownership issues and volatility of cash flows due to demand risks are key. Need to address volatility, risk management, and financing costs, but these can have adverse incentive effects	Positive cash flows and risk of default substantially diminished	Refinancing of debt from the initial phases with possibility of bond issue, bank loans, or infusion of government funds

Source: Torsten Ehlers. Understanding the challenges of infrastructure finance. 2014.

will carry the greatest exposure to such risks. Debt holders will expect to see an equity contribution, plus be satisfied as to the specialized skills and technical expertise of the contactor and the monitoring devices put in place. Investors must rely on this expertise as they will not have the requisite monitoring capabilities. This makes large construction firms possible candidates as equity contributors. With their money at risk, they will focus on risk management measures to address changes in design and scope, construction delays, and cost increases. A government sponsor may want to retain a share in a project and contribute to some of the equity requirements at the construction stage. This could be funded from a capital budget. Bond financing at the construction stage is the least desirable as it offers limited or no flexibility if existing loans must be renegotiated for whatever reasons.

It is at the operations and maintenance stage, when revenues are generated and final costs are established, that governments have the best chance of attracting private capital. Financing at this stage may require de-risking strategies, credit guarantees, or credit insurance that can be purchased privately or underwritten by the public entity. An example of de-risking would be a floor on revenues below which government would share revenue risks. This is when project finance syndicated loans, or single-bank loans can be used, as well as issuing bonds. With stable underlying cash flows, projects mirror fixed income securities and become attractive as a low-cost way to raise funds to retire debt. This assumes the existence of a market for these bonds. The infrastructure bond market is still relatively small compared to the syndicated loan market. Investors are generally not attracted to low-risk/low-return bond yields that may fall well below their investment threshold.

The infrastructure bond market

Worldwide, infrastructure investors are increasingly tapping into bond markets to refinance existing assets and expand the range of investment opportunities, including the use of green bonds (Lam-Frendo, 2021). Although bonds represented only about 20 percent of private infrastructure financing in primary transactions in developing countries, as of 2020, this has nearly doubled since 2015. Internationally, infrastructure bonds include municipal special-purpose bonds and corporate infrastructure bonds[3]. The most developed market for such bonds exists in the United States, India, Australia, Chile, Canada, and Kazakhstan. The use of bonds varies widely across countries but significant differences exist. The contrast between Canada and the U.S. is striking. In the U.S., the $3.8 trillion municipal bond market sees debt issuance exceed $400 billion each year. Although Canada is 10 percent the size of the U.S., based on population, the Canadian municipal bond market is less than 1 percent of the size of the corresponding U.S. marketplace, or a mere C$35 billion. Canada sees less than C$5 billion in municipal debenture issues annually (Fenn, 2022).

Revenue bonds refer to a type of U.S. municipal bond issued for a specific project and secured by a source of revenue as a result of the project's implementation. An example of such projects is the construction of a toll road or airport. These types of bonds are most often issued by state and local government agencies responsible for specific types of operations. Unlike general obligation bonds, the coupons and principal on revenue bonds cannot be paid out of the tax revenues of the issuer. If the cash flows generated by the project are insufficient to service the bonds, the investor is likely to be short of income.

In addition, investors in revenue bonds may not claim the assets of the project in the event of insolvency. Thus, the reliability of this type of bond is mainly determined not by the issuer's creditworthiness, but by the profitability of the project. Accordingly, as a rule, the revenue bonds have a higher return than the general obligation bonds with similar parameters.

Bonds have been used in emerging markets but there are constraints. They are normally issued in local currencies to minimize currency mismatches, and a local bond market is required[4]. There is also the need for a sound legal framework, bureaucratic efficiency, and contract enforceability. Bonds must also be rated to be attractive to investors and several bond initiatives have been launched including India and China. In China, bonds have been issued entirely by state-owned enterprises (SOEs) which imply government guarantees. In other Asian countries bonds have usually always been issued by a separate project SPV, which retains the incentive to push for the successful execution of a project. Back in the early 2000s, India adopted several policies and a legal framework to facilitate urban local government access to capital markets through municipal bond issues. Although limited in use, these bonds offer alternative modes of financing for urban infrastructure (Chattopadhyay, 2006).

Infrastructure bonds in India are usually issued by banks[5]. Two types of infrastructure bonds are common: tax-saving bonds and regular income bonds. Tax-saving bonds give the right to receive a deduction from the tax base for income tax purposes, equivalent to 20 percent of the investment amount. Regular income bonds are represented by pension, educational, and other bonds. The disadvantage of the Indian infrastructure bond market is the lack of protection against inflation. Infrastructure bonds in India account for about 25 percent of issuances in the overall corporate bond markets (Indian Infrastructure, 2021). Public sector entities, such as the National Highways Authority of India (NHAI), NTPC Limited, the National Hydroelectric Power Corporation, and Power Grid Corporation of India account for almost 50 per cent of infrastructure bond issuances, whereas private sector issuances are diversified across issuers and sectors.

In recent years, more than 90 per cent of infrastructure bond issuances in India have been from NHAI in the road sector. Traffic volume risk and sponsor risk continue to be key reasons for risk aversion on the part of investors. Issuances from the private sector are relatively better in the power sector, accounting for 50–55 per cent of issuances. The encouraging factors include presence of relatively strong sponsors, cross-default structures across various SPVs, and relatively few operational risks. The renewable energy sector accounted for around 40 per cent of the overall bond issuances from the power sector.

Interest in sustainable finance is growing across global capital markets (Figure 14.3), giving rise to different bonds including green, social, blue, and sustainable bonds (Chase, 2021). These are proceeds bonds, the uses of which are used to finance or refinance eligible infrastructure projects, or assets that fall within specific categories.

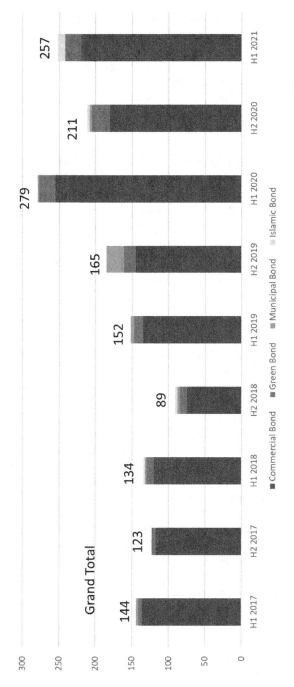

Figure 14.3 Global bonds market for infrastructure finance (2017–2021).

Source: GIHub 2021.

- **Green bonds:** Funds are committed to environmental or climate projects, such as investing in renewable energy.
- **Social bonds:** Funds are committed to social impact projects, such as investing in low-cost housing for people with restricted access to the housing market.
- **Blue bonds:** Funds are committed to marine or water projects, such as investing in the transition to sustainable fish stock.
- **Sustainability bonds:** Funds are committed to a mix of social and green impact projects. These projects may also be aligned with the UN Sustainable Development Goals (SDG).

A key component of sustainability-linked bond (SLB) agreements is the linking of the financial terms to the achievement of predetermined sustainability performance targets. These forms of sustainability finance will continue to play a significant role in tackling environmental and social issues, driven by demands from regulatory and industry bodies, investors, shareholders, and those in pursuit of investment strategies to meet their ESG targets.

Lowering the cost of capital: The Ontario experience

All governments will pursue the lowest cost of capital in their infrastructure investments, as they should. However, this can also interfere with the ability to attract private capital in the absence of a well-thought-out strategy that can accommodate the interests of both sides. The experience of Ontario, Canada, in the use of PPPs for both social and economic infrastructure illustrates the importance of strategic thinking, particularly when capital markets are under stress.

Ontario, Canada, has enjoyed a robust market for PPPs for more than a decade and involving both public funds and short- and long-term private financing, administered by a single government agency, Infrastructure Ontario (I/O). This agency is credited with advancing the PPP model to a level at which it has been recognized as a benchmark for best practices. The need for I/O to reengineer its investment processes in support of PPPs was prompted by the Great Financial Crisis (GFC) of 2008, following which commercial lenders (banks) retreated from the infrastructure market and borrowing costs went up. At the height of the crisis, a particular European bank refused to uphold its commitment for construction lending for a PPP hospital project sponsored by Infrastructure Ontario. Niagara Health would be the first PPP sponsored by Infrastructure Ontario that failed to reach commercial close. This was not deemed acceptable from the perspective of the market and the reputation of Infrastructure Ontario.

In response, the Government of Ontario instituted a series of temporary measures, including a schedule of advance payments during construction to reduce the cost of a commercial construction loan; payment of 50 percent of the total project cost to the private partner at time of substantial completion;

and the remaining 50 percent paid out to the private partner over the term of the concession. Infrastructure Ontario took it upon themselves to assembling a group of Canadian banks to syndicate the construction loan and underwrite a bond issue to finance the deal. Most of what were considered interim measures were subsequently adopted as standard practice in PPPs subsequently issued by Infrastructure Ontario. The Government of Ontario announced they would pay down 60 percent of a total project cost at substantial completion for all social infrastructure projects, and up to 85 percent for transportation projects. This reflected their lowering borrowing costs and unrealized borrowing capacity. This injection of public capital right up-front reduced long-term financing costs over the term of the concession, while retaining sufficient leverage over the private service provider (SPV) to maintain high-quality services during the concession period.

These new measures had several positive effects. First, there was now an option to traditional bank lending for construction and driving down the cost of construction debt. Second, the arrangement offered construction companies a greater role in PPPs with the opportunity for equity participation in lieu of traditional equity sources. Third, this initiative gave rise to an infrastructure bond market, with both short- and long-term bonds that could combine short-term construction lending with long-term financing. Fourth, for construction lending, hesitant commercial banks could participate in what were called "club deals", with a lead bank syndicating a loan to other banks to spread the risk and thereby lower the interest premium. Finally, there was the opportunity for investors to buy project bonds which had direct appeal to large Canadian insurance companies. In a very challenging financial environment following the GFC, these initiatives aligned government objectives with industry needs by offering incentives to better manage risks and meet performance obligations.

From this experience, Infrastructure Ontario realized that it could tap private capital in the form of commercial bonds that carried a public guarantee derived from the availability payments. This brought private lending costs more in line with the cost of government debt. Private capital in the form of a very small equity contribution by the SPV would give government leverage over the operator during the concession period and motivate the operator to meet performance requirements. This arrangement gave immense oversight powers to the bond holders who would want to avoid any chance of default, or the need to trigger their "step-in" provisions. A large but unquantifiable residual risk still remained for government. What if the private partner realized that they took on too much risk at too low a price, and the solution was to "walk" from the deal. This is particularly true of large complex multi-billion-dollar projects where mispricing of risk during construction, and over the term of the concession agreement, can lead to default.

For government, private debt compared to private equity was much cheaper, with equity requiring a return perhaps more than twice that of debt. Unfortunately, this is the opposite to what most private investors seek. Private investors, and particularly institutional investors, seek large equity positions and minimal debt. They prefer large transactions, fewer deals to keep costs down, and avoid risks they are unfamiliar with. In comparison to commercial infrastructure deals, PPPs have not proven to be an attractive proposition for private investors in infrastructure, even when investing in developed countries.

Case study: Bridgepoint Health Care, Toronto, Canada

Bridgepoint Health Care was the second PPP hospital that Infrastructure Ontario was involved with following the GFC, and the first to formally implement the new policies and procedures that were introduced to achieve financial close with Niagara Health.

The resulting capital structure of Bridgepoint is depicted in Table 14.2. Three features stand out. First, the equity contribution is very small, only 7 percent of the total capital cost and an amount that a large contractor could raise with a small, syndicated investment fund. Second, 49 percent of the capital cost is financed by private bonds. Third, government funds 44 percent of the capital cost with a single payment at substantial completion. This payment covers 67 percent of the design and construction costs and was sufficient to retire the construction loan which carried a high interest rate. From a private capital perspective, there must be a private market for these bonds, or this approach fails. The large institutional investors have little or no interest in PPP bonds as the can buy commercial bonds that offer much more liquidity. However, there was a strong market for these bonds among Canadian insurance companies who saw infrastructure bonds as low risk and aligned with their long-term cash

Table 14.2 Breakdown of Financing Sources, Bridgepoint Health Care

Plenary Health Bridgepoint L.P. Sources and Use of Funds (Consolidated)

Source of Funds	Mil. C$	Use of Funds	Mil. C$
Senior bonds	252.1	Design and construction costs	333.4
Substantial completion	225.0	Interest and commitment fees	82.1
Equity contributions	33.8	Senior Debt Service Reserve Account	10.5
Interest received/other	4.9	SPV and operating costs	19.1
		Development and financing costs	19.4
		Equipment cash allowance	6.7
		Construction period cash account	44.2
		Other fees and expenses	0.4
TOTAL SOURCES AVAILABLE	**C$515.8 M**	**TOTAL**	**C$515.8 M**

Source: Hussain and McKellar, 2020.

requirements. A new source of private capital entered the PPP infrastructure market in Canada.

Humber River Hospital was the next PPP hospital to be built after Bridgepoint and with the success of the bond issue for Bridgepoint, the developer pursued a financing solution that would entirely eliminate the need for a commercial construction loan. Construction would be financed with bonds using a mix of short- and long-term bonds. Bond purchasers were required to purchase a combination of both, in a predetermined ratio to cover up-front costs. The short-term bonds had a 7-year expiry date and were paid off largely from the proceeds of the substantial completion payment from government. The overall effect was a reduction in the cost of financing during construction for what is normally the highest price debt.

Notes

1 The FDN, sourced at www.fdn.com.co
2 World Bank, through its Project Preparation Facility (PPF), may make an advance from the PPF to a borrowing country to finance preparatory activities for investment operations, including preliminary and detailed designs and limited initial implementation activities, and preparation of programs to be supported by development policy lending operations. The PPF complements other bank methods of assisting project preparation, such as emergency recovery assistance loans, technical assistance loans, grants, and retroactive financing[2]. It is used only when these methods are not suitable and alternative sources of funds are not available. A PPF advance is made only to a member country.
3 CBonds. Infrastructure Bonds. https://cbonds.com/glossary/infrastructure-bonds/
4 CBonds. Infrastructure Bonds. https://cbonds.com/glossary/infrastructure-bonds/, p. 19.
5 CBonds. Infrastructure Bonds. https://cbonds.com/glossary/infrastructure-bonds/

Part IV
Sustainable infrastructure

15 What is sustainable infrastructure

Anything including the word "sustainability" might be suspect given that "greenwashing" is a widespread marketing technique. In the public realm, the term sustainability appears in government policy documents, new government initiatives and programs, and in political rhetoric. What then is "sustainable infrastructure", particularly from an investment perspective?

> According to a report from Brookings Institution (Bhattacharya et al., 2016), 2015 was a milestone year in which the world set clear and ambitious objectives through the Third International Conference on Financing for Development in Addis in July; the UN Summit in September that adopted the Sustainable Development Goals and the 2030 development agenda; and the COP21 in Paris in December that resulted in the milestone climate agreement. The three central challenges now facing the global community, as crystallized in 2015, are to reignite global growth, deliver on the sustainable development goals (SDGs), and invest in the future of the planet through strong climate action. At the heart of this new global agenda is the imperative to invest in sustainable infrastructure.

In much of the literature the difference between spending on infrastructure, versus spending on sustainable infrastructure is not clear. Questions arise. How do you draw distinctions between infrastructure and sustainable infrastructure? Are both addressing the same needs or does sustainable infrastructure address a broader spectrum of needs? Does sustainable infrastructure perform differently from standard infrastructure? Does it cost more?

In the absence of a universally acknowledged definition of sustainable infrastructure, there is room for misunderstanding over what the term means to the private investor versus a public official, or even a user. Will investors begin to shift preferences to sustainable infrastructure and, if so, why? There is also a debate over ESG and its role in promoting sustainability.

DOI: 10.1201/9781003396949-20

What is sustainable infrastructure and why do we need it?

Answers to these questions depend on who you ask. There appears to be a shift in the investment industry toward more sustainable projects, whether this is driven by recognition of the impacts of climate change, government policies and commitments, or market pressures. Responses will increasingly reflect a definition of sustainable infrastructure that refers to infrastructure that is socially inclusive, low carbon, and climate resilient (Bielenberg et al., 2016). Nor is there much debate, other than scattered political opposition and comments from climate-change naysayers, on the need for sustainable infrastructure. Responses on the public side embrace commitments to the United Nation's Sustainable Development Goals (SDG). At the first Conference of Parties (COP) held in Paris in 2015, almost 190 countries, representing more than 98 percent of greenhouse gas emitters, agreed to a global climate-change strategy and a commitment to a temperature rise limited to 1.5°C. The 2021 COP26 in Glasgow, Scotland, the fifth since Paris, confirmed how slow progress toward SDG has become (Economist, 2021). COP27 held in Cairo, Egypt, in 2022 raised the question as to whether this assembly can move the needle in the right direction. Since 2015, the severe additional impacts of warming at 2°C, rather than the lower Paris limit, have become increasingly clear. COP27 indicated that momentum for change is building, despite intense lobbying from fossil fuel interests. Two new initiatives stand out: acknowledging the need for funding for loss and damage by the richer nations; and finally placing food and agriculture under scrutiny.

On the investment side, major institutional investors are pledging to decarbonize their investment portfolios and assess the carbon footprint of their assets. Like the public sector, these initiatives will encounter stubborn resistance as evidenced by the strong lobby presence from the oil and gas industry at COP27[1]. However, progress toward decarbonizations is slowly picking up pace, particularly with pressures from within the investment industry to meet ESG requirements. Shifting entire portfolios to meet sustainability objectives affects not only infrastructure, but other alternative assets and equity investments. This shift is an essential strategic exercise driven top-down and through governance mandates. This is a slow, complex, and relatively costly task for any institutional investor who must maintain a fine balance between return expectations, addressing the interests of stakeholders, and being environmentally responsible[2].

Most greenfield projects will involve some form of public–private collaboration, and investors will be beholden to governments to launch these infrastructure projects and specify their sustainability objectives. The pace, structure, and pipeline of projects launched by governments will be a major factor in whether investors can reach their sustainability targets for new projects.

Sustainability is context sensitive

Sustainability is context sensitive. An example would be building a new toll road connecting two urban centers in an emerging economy, compared to building an almost identical toll road to address traffic congestion in a growing metropolis in a developed country. One could argue that the toll road in a developing economy has significant social and economic benefits, even though gasoline-powered cars and trucks will exacerbate air pollution problems. It is a solution to mobility likely to prevail for many years as emerging economies struggle to advance their economies and pull people out of poverty. Those opposing a toll road in a developed country could argue that it is not a solution to traffic congestion, contributes to greenhouse gas emissions, has significant impacts on the local flora and fauna, and is a solution well past its time. There is no perfect solution.

Where there is general agreement is that infrastructure is an essential foundation for achieving inclusive growth and expanding economic activity. It is an essential ingredient to boosting human capital and quality of life through its positive effects on food security, the provision of clean water and sanitation, and the delivery of health care and education. There is also a shared recognition of the negative effects brought about by some types of infrastructure. The negative impacts include air and water pollution, release of other toxic pollutants, irreversible damage to the natural landscape, consuming non-renewable resources at a non-sustainable pace, and prolonging practices that contribute to environmental degradation such as clear cutting of timber, construction of dams, and reliance on concrete as a primary building material. Despite knowing that infrastructure choices exercised today may compromise quality of life for future generations and create unsustainable economic burdens in the future, society has a habit of continuing down the same path for as long as possible.

There are attempts to mitigate business-as-usual trends which include purchasing carbon credits from developing countries, utilized as a tool to offset carbon and associated damages created by developed nations. New promising technologies include the creation of carbon capture concrete, large-scale carbon sequestration facilities in the form of carbon capture utilization and storage (CCUS) technology, hydrogen energy, and other emerging technologies which will present their own environmental challenges as they develop into more viable solutions. All this new technology will face context-sensitive challenges.

A narrow window of opportunity

Embarking upon a commitment to sustainable infrastructure has the expectation that this will accelerate the transition to an economy based on low-carbon energy, capitalize on investments that can utilize new low-impact technologies, and promote patterns of economic activity that can contribute to improved quality of life. The transition to developing sustainable infrastructure is particularly timely given the global macroeconomic context, the early signs of a slowdown in economic growth, and dealing with the consequences of the COVID pandemic and the situation in Ukraine, consequences which will manifest in the coming years. Unfortunately, the benefits of a shift to sustainable infrastructure will not be immediate, nor obvious, will be costly in the interim and, as with all transformational changes, perceived benefits will take time to materialize.

In the short to medium term, sustainable infrastructure can improve the efficiency of energy utilization, mobility, and logistics, thereby boosting global productivity and competitiveness among nations. But this can only happen if ways are found to engage large amounts of private capital, involve the private sector in new delivery models, and unlock private innovation and creativity. Those who believe that the public sector can go it alone must face the stark reality of fiscal policy constraints across most governments, and the real possibility of economic contractions following years of exuberant growth, fueled by low interest rates and the benefits of global supply chains.

While the benefits of sustainable infrastructure are long term, the window through which to launch change is very small, measured in years, not decades. Within this relatively narrow window of opportunity, three things must happen (Bhattacharya et al., 2016).

- First, in developed countries, large investments must be directed at rehabilitating the stock of existing infrastructure that has long suffered from under-investment. Rehabilitation should normally take priority over new infrastructure, although this will not sit well among some politicians who like cutting ribbons for new projects. Rehabilitation initiatives may conclude that some existing infrastructure is functionally obsolete, or in some situations prove too costly an undertaking.
- Second, the focus must shift to the emerging and frontier economies which have higher growth rates, constitute a higher share of the global economy, and have the greatest needs driven by rapid urbanization, changes in economic structure as they increase their presence in global markets, and have increasing populations and a growing middle class.
- Third, increasing investment in infrastructure in the emerging economies must be sustainable given the magnitude of investment activity anticipated and the impact this will have on achieving low-carbon scenarios and addressing climate resiliency. This implies investments in renewable energy, nuclear, carbon capture storage, transportation, water and sanitation, and adaptive infrastructure to better withstand climate change impacts.

Barriers that inhibit the flow of funds into infrastructure that exist across countries are magnified when it comes to committing to sustainable infrastructure. These barriers make the task even more challenging in emerging economies that lack coherent legal and institutional frameworks, limited capacity to deliver projects, poor governance models, and government-induced policy risks that affect both the delivery and operation of infrastructure projects. It must also be acknowledged that sustainable infrastructure projects, no matter where located, will cost more. McKinsey estimates that infrastructure that meets sustainability requirements will likely increase up-front capital costs by 6 percent or more for individual projects (Bielenberg et al., 2016). Higher costs translate into higher financing barriers.

Significant impediments to shifting to sustainable infrastructure projects are the price distortions in the pricing of natural resources and infrastructure services including fossil fuel subsidies and lack of carbon pricing (Bhattacharya et al., 2016). These distortions favor high-carbon energy sources and fail to price the damage from pollutants that they spew into the environment. Price distortions also discourage the development of cleaner energy technologies and fail to account for the impact they have on the public purse both on the costs and revenue side. According to the Brookings study,

> The IMF has estimated that the total cost of energy subsidies, including the failure to price in negative externalities in terms of pollution and climate change impacts, was $5.3 trillion in 2015, or 6.5 percent of world GDP. Elimination of fossil-fuel subsidies would reduce global CO_2 emissions by more than 20 percent, cut premature deaths from air pollution by more than half, and could generate substantial fiscal gains of $2.9 trillion (3.6 percent of world GDP in 2015).

Sourcing the necessary capital

The public sector will maintain a central role in guiding new infrastructure investments given the extent of their direct involvement in infrastructure and their policy and regulatory roles in affecting any private sector involvement in these initiatives. They can also play a crucial role in incentive programs to direct investors toward sustainable projects in lieu of carbon-intensive projects. Their own budget allocations have historically been the major source for infrastructure financing, particularly in developing countries. Where the need is greatest, at the municipal level, funding sources are scarce and primarily rely upon capital budgets funded from local tax revenues. Financing options at this level are very limited and local governments will need to consider the private sector in launching new sustainability initiatives.

A most pressing challenge will be the conversion of brownfield infrastructure to meet sustainability targets. Transportation is a good example where diesel fuel is widely used in buses, commercial vehicles, and locomotives. Climate change is already a key driver in the need to adapt existing infrastructure to address the dire consequences of rising temperatures. The obvious solution in transportation is conversion to electricity or even hydrogen. Should this occur on a large scale, it will likely be funded either by corporate owners of infrastructure or by private investors, as public coffers will be severely constrained.

Can this happen in the absence of a clear and comprehensive national strategy to achieve climate goals? A clear and coherent public policy framework is required that can provide the clarity and confidence to private investors so they can do their part. It may also require government support through financial incentives, changes in regulations, new financing instruments, and risk-sharing agreements. In addition to mobilizing the private sector, public investment must be ramped up to meet projected growth in demand and address challenges of sustainability for new infrastructure projects.

The good news is that the overall volume of private investment has grown rapidly over the past two decades. The bad news is that, with a few exceptions, very little of this capital is directed to long-term public investments, and even less is being made available for infrastructure financing. Despite governments advocating an increasing role for private capital to fund infrastructure, there is little evidence of progress on this front. In developing countries, attracting private capital has been on a downward trend over the past decade.

A dirty secret

The dirty secret according to the Economist, is that "the construction industry remains horribly climate unfriendly" (Economist, 2022). Most infrastructure projects involve large-scale construction which is plagued by two problems: poor productivity, and perhaps more significant, new construction and associated demolition which are among the planet's worst climate offenders. Construction involves massive amounts of steel and cement, both of which are huge contributors to CO_2 emissions (Figure 15.1). Demolition releases "embodied" carbon, which refers to emissions tied to the building process, as well as maintenance and any demolition. Embodied carbon is responsible for around 10 percent of annual global emissions and will vary depending on the type of structure (Economist, 2022).

The biggest culprit is concrete, the most widely used man-made material in existence, second only to water as the most-consumed resource on the planet. Cement is the source of about 8 percent of the world's carbon dioxide emissions according to the thinktank Chatham House (Rogers, 2018). If the cement industry were a country, it would be the third largest emitter in the world, behind China and the U.S. Cement contributes more CO_2 than aviation (2.5 percent) and is not far behind the global agriculture business (12 percent) (Figure 15.2).

Addressing the cement problem is a complex matter given the anticipated construction frenzy in the decades ahead (Czigler et al., 2020). There will be the impacts of growth and urbanization in China as well as India, continuation of outdated practices of the construction industry, and the penchant for large engineering marvels built of concrete that continue unabated. There are no clear solutions, but several initiatives are proposed.

- Begin to implement performance standards and regulations that apply to both new structures and retrofits.

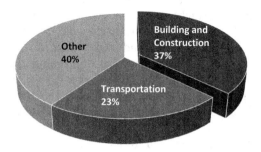

Figure 15.1 Energy-related CO_2 emissions, 2020 (% of global industry total).

Source: Economist, 2022.

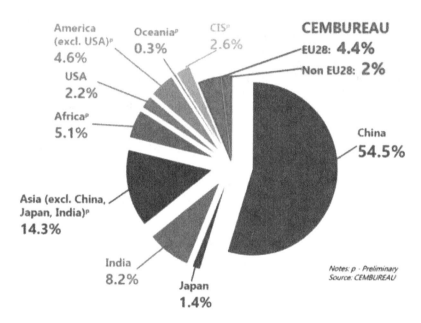

Figure 15.2 China produces most cement and therefore most-cement related CO_2 emissions.

Source: CEMBUREAU, the European Cement Association.

- Mobilize to restrict the wholesale demolition of existing buildings that release embodied carbons. Land use regulations and even tax codes are skewed to favor replacing old buildings with new structures with minimal effort to have replacements meet green standards. New structure should be subject to whole-life carbon assessments with net-zero building code requirements.

- Explore alternatives to traditional concrete and new negative carbon concrete technology. This can be augmented with new renewable energy sources and potentially carbon-negative solutions.
- Explore new building practices and carbon-neutral materials and technologies such as the use of heavy timber and timber composites for low- and medium-rise buildings and explore opportunities for zero-waste prefabricated assemblies. Sustainable infrastructure must be an integral part of the search for net-zero emissions and potentially carbon-negative solutions.
- Repurpose captured carbon. Reintroduce CO_2 emissions captured from the production processes into the value chain. Carbon capture, utilization, and storage (CCUS) represents the biggest decarbonization lever in the path toward net zero and offers a massive opportunity for innovation.

Notes

1 There were more than 600 fossil fuel lobbyists at COP27, a rise of more than 25 percent from the previous year and outnumbering the community affected by the climate crisis. www.theguardian.com
2 The intense debate and media attention over Blackstone's recent issuing of their policy initiative, "An Integrated Approach to ESG", is an example of the main sides and stakeholders involved in this debate. Sourced at www.blackstone.com/our-impact/an-integrated-approach-to-esg/

16 Global landscape of climate finance

According to the United Nations (UNFCCC, 2022):

> Climate Finance refers to local, national or transnational financing—drawn from public, private and alternative sources of financing—that seeks to support mitigation and adaptation actions that will address climate change.

Climate change creates climate-related risks that are incorporated in several different types of risk. *Transition risks* involve changes in law, policy, technology, and markets related to the transition to a lower-carbon energy supply (Center for Climate and Energy Solutions, 2022). Those risks include regulatory risk such as climate laws and policies that affect how companies operate, and liability risk, such as litigation that targets companies for contributing to climate change and affects a company's reputation. *Physical risks* from climate change include damage to fixed assets like buildings and property, or supply chain disruptions. Damage can result from extreme weather events or changes in water availability. The effects of climate change are not all negative. Climate change can also present opportunities, including expanded markets for existing products and services, new markets for new and existing products, and cost reductions.

The 2021 edition of the "Global Landscape of Climate Finance 2021" offers a definitive guide on this topic (Climate Policy Initiative, 2021). Figure 16.1 provides a comprehensive but complex overview of the labyrinth of capital flows from sources of capital through sectors funded by climate finance initiatives. Overall, the sources of capital are many and quite diverse, including both public and private sources. Of the $632 billion, the public sector accounts for 51 percent of tracked climate finance, of which the development finance institutions (DFIs) contribute the majority (70 percent). On the private side, commercial financial institutions and corporations together contribute almost 80 percent of the total. Climate finance requires grants, of which governments are the main source of grant funding. Most climate finance in 2019 and 2020 was raised as debt, of which 88 percent

DOI: 10.1201/9781003396949-21

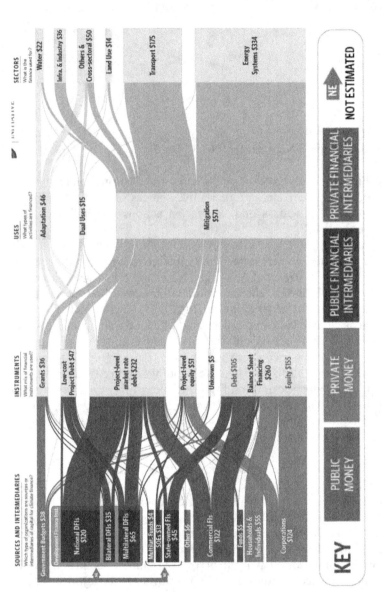

Figure 16.1 Global climate finance flows in 2019/2020.

Global climate finance flows along their life cycle in 2019 and 2020. Values are average of two years' data in USD (632 Bn USD annual average)

Source: Climate Policy Initiative, 2021.

was market-rate debt. Concessional, low-cost project debt (12 percent) was provided by public institutions. Equity investments represented 33 percent of total climate finance.

The use of this climate-related capital is highly focused on mitigation measures with a relatively small amount committed to adaptation. The primary recipient of much of this capital is in the energy sector (53 percent), followed by transportation (28 percent). Other sectors receive a relatively small amount. Solar PV and onshore wind were the main recipients of renewable energy finance and low-carbon transport is the fastest growing sector.

More than 76 percent of climate-related investments flowed domestically—raised and spent within the same country, with three-quarters of these investments concentrated in East Asia and the Pacific region with China the dominant recipient, followed by Western Europe, and North America. In economically advanced regions, investments were primarily from the private sector.

According to the report by the Policy Climate Initiative, total climate finance has steadily increased over the past decade, growing from $364 billion in 2011/2012 to $632 billion in 2019/2020[1]. Whereas the rate of increase averaged 25 percent annually starting in 2013/2014, it slowed to 10 percent after 2017/2018. But even at a 25 percent growth rate, this palls in comparison to the 590 percent annual increase that is estimated to meet internationally agreed upon climate objectives by 2030. Adaptation financing is a particularly weak performer with almost all adaptation financing coming from the public sector[2]. This performance may reflect the fact that data on adaptation financing carried out by the private sector is largely missing.

Adaptation measures to address climate change are very limited and involve practices that can range from crop diversification, new irrigation systems, water management, disaster risk management, to risk insurance. Some of these initiatives may be combined with social and economic goals. They are undertaken by a range of public and private actors through policies, investments in infrastructure and technologies, and incentives to drive behavioral changes. Some adaptations represent incremental shifts from the status quo such as influencing consumer choices, and some are transformative such as promoting the electric car as an alternative to ICE vehicles, or the use of solar panels as a clean energy source for a house. Adaptations in higher-income countries can be directed at influencing personal or household choices with relatively low-cost solutions, whereas in low-income countries the greatest need is for sustainable infrastructure in the energy and transportation sectors involving large capital expenditures (Nature, 2021).

Public climate finance remains relatively stable at around $321 billion or 51 percent of the total. Most of this money (69 precent) comes from DFIs, with smaller amounts from state-owned financial institutions (14 percent) and direct government grants (12 percent). On the private side, corporate spending accounts for the largest share of private climate finance (40 percent), followed by commercial financial institutions, and household spending driven by annual consumer spending. Most of climate finance (61 percent) is raised as debt, of which 12 percent is low-cost or concessional debt.

Solar PV and onshore wind are the largest recipients of renewable energy finance, mainly with private capital, indicating the commercial viability of this sector. The fastest growing sector is low-carbon transportation (battery electric vehicles) reflecting a combination of government subsidies, falling technology costs, and evolving consumer markets. Investments in buildings and infrastructure remain low but also the data are hard to collect and subject to agreeing on what constitutes infrastructure at a building level.

There are clear geographic distinctions in where the money originates from and flows to. More than 75 percent of 2019/2020 tracked climate investments flowed domestically, monies raised and spent in the same country.

The climate financing gap

As Figure 16.2 indicates, climate finance flows are nowhere near estimated needs, conservatively estimated at $4.5–5 trillion annually. To achieve the transition to a sustainable, net zero emissions, and resilient world in the next decade, climate-related investments must increase drastically.

In addition to the flows of capital depicted in Figure 16.2, there has been significant action taken to provide information to investors about what companies are doing to mitigate the risks of climate change. At the forefront of this initiative is the Task Force on Climate-related Financial Disclosures (TCFD).

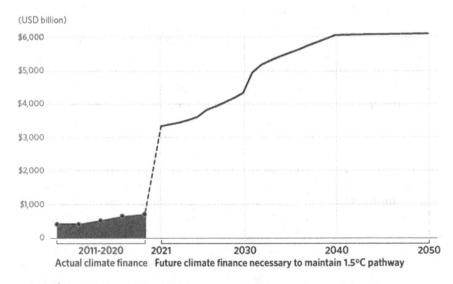

Figure 16.2 Global tracked climate finance flows and average estimated annual climate investment needed through 2050.

Source: Climate Policy Initiative (CPI). 2021. Global Landscape of Climate Finance 2021.

Task Force on Climate-related Financial Disclosures (TCFD)

The TCFD was established in 2015 by the G20 Financial Stability Board to develop recommendations on the types of information that companies should disclose to support investors, lenders, and insurance underwriters in assessing and pricing a specific set of risks—risks related to climate change (TCFD, 2022). Disclosure is structured around four thematic areas that represent the core elements of a company's operations: governance, strategy, risk management, and metrics and targets. As of January 2022, the number of TCFD supporters exceeded 3,000 organizations in 92 countries, with a combined market capitalization of $27 trillion. Another organization active in climate-related investments is the Coalition for Climate Resilient Investments.

Coalition for Climate Resilient Investments (CCRI)

Launched at the UN Secretary General's Climate Action Summit in 2019, with the support of the UK Government, the Coalition for Climate Resilient Investment (CCRI) is a private-sector-led initiative dedicated to supporting investors and governments to better understand and manage physical climate risks (CCRI, 2022). The mission of CCRI is to mobilize private finance, in partnership with key public institutions, to develop and implement practical solutions for the effective integration of PCRs in investment decisions and the acceleration of investment flows in climate resilience. CCRI has a membership of over 120 institutions, covering the entire financial and physical infrastructure ecosystem and including international convening partners and organizations, governments, multilaterals, non-profits, think tanks and academics, institutional investors, asset managers, pension funds, banks, insurers, standard setters, ratings agencies, lawyers, engineers and developers, consultants, auditors, and financial and climate data providers. As of April 2022, financial sector participants represent over $25tn of assets.

Notes

1 The terms climate finance and climate investment are used interchangeably.
2 The term adaptation financing refers to projects directed at reducing a country's and community's vulnerability to climate change by increasing their ability to absorb impacts and remain resilient.

17 ESG and responsible investing

A brief history of environmental, social, and governance (ESG)

According to Forbes, the beginning of ESG dates back to January 2004 when former UN Secretary General Kofi Annan wrote to over 50 CEOs of major financial institutions inviting them to participate in a joint initiative under the auspices of the UN Global Compact and with support from the International Finance Corporation (IFC) and the Swiss Government. The goal was to find ways to integrate ESG into the capital markets. Following this meeting a report was released in December 2004 entitled "*Who Cares Wins*", authored by Ivo Knoepfel (Knoepfel, 2004). The proceedings of the original conference were published in 2005 (International Finance Corporation, 2005). The case was made that embedding environmental, social, and governance factors in capital markets makes good business sense and leads to more sustainable markets and better outcomes for society.

These two reports formed the backbone for the launch of the Principles for Responsible Investment (PRI) and the launch of the Sustainable Stock Exchange Initiative (SSEI) the following year. Forbes estimated back in 2018 that ESG investing was over $20 trillion in AUM, or around a quarter of all professionally managed assets around the world (Kell, 2018). Unlike its predecessor, socially responsible investing (SRI) which has been around much longer and based on ethical and moral criteria, ESG investing assumes that ESG factors have financial relevance. ESG investing accelerated around 2013 and 2014 when early research indicated that good corporate sustainability performance correlated with good financial results. In a short period ESG has grown to become a very large business, but its rise has been neither smooth nor linear (Kell, 2018). However, it is no longer recognized as a fad and is regarded as an indication for how markets and society have progressed. The presence of ESG considerations confirmed that methodologies for asset valuation were recognizing the importance of sustainability factors.

Environmental, social, and governance issues, under the acronym ESG, still remain controversial, which attests to their relative newness, the lack of sufficient evidence to justify the claimed benefits, and lack of clarity in just what is driving the commitment at embrace ESG among private capital providers.

DOI: 10.1201/9781003396949-22

Yet, there is almost universal appeal among money managers to demonstrate, through a commitment to ESG, their sensitivity to and progress toward better environmental management. Industry leaders will use arguments in favor of ESG that range from it being an ethical imperative to believing that it creates long-term value. According to the Economist, ESG is "an attempt to make capitalism work better and deal with the grave threat posed by climate change. It has ballooned in recent years; the titans of investment management claim that more than a third of their assets, or $35trn in total, are monitored through one lens or another. It is on the lips of bosses and officials everywhere" (Economist, 2022).

Shifting the emphasis in ESG

ESG involves both financial and non-financial factors, matters of quality and quantity, and an acknowledgment that climate change and transition to a low-carbon economy will affect the future performance of an asset. Of the three letters, environment—the "E" in ESG—is perhaps the easiest to define and deal with. Social and governance issues are much more challenging to define and more controversial as they cover such a broad array of factors. These can range from health and safety to community acceptance, and internal company practices that may involve policies on diversity and inclusion, business performance, and leadership attributes. It is easier to measure carbon emissions than social impact. It is much easier to tick off the "E" box than the "S" and "G" boxes. In fact, the Economist makes the case that the three letters: E, S, and G should be unbundled as they represent too many targets to hit and reduce the "chance of bullseye-ing any of them" (Economist, 2022). Their claim is that, in the face of the threats from climate change, ESG should be boiled down to one simple measure: emissions.

Challenges in implementing ESG

It must be acknowledged that there are wide variations in interpreting ESG across mature markets, between brownfield and greenfield projects, in disclosure and reporting requirements, and involving different emphases in applying ESG practices. Europe has a strong focus on environment and climate change, whereas in the U.S. there is greater emphasis on workers' rights, working conditions, and workplace safety (Stanley, 2019). Australia tries to encompass both ends of this spectrum. Applying ESG in emerging markets brings into focus challenges with corruption, bribery, worker protection, and safety. ESG casts a wide net (Figure 17.1).

A large part of the controversy surrounding ESG stems from the lack of quality and consistent environment and sustainability data. In a survey by Preqin of unlisted fund managers, 71 percent cited this as a challenge in implementing ESG, followed by confusion over industry terminology

Figure 17.1 Infrastructure fund manager views on the main challenges of implementing ESG policy.

(57 percent) (Preqin, 2021). While regulatory and legal requirements for compliance are a key driver of an increased focus on ESG, there should also be enhanced financial performance, and this is where the debate gets murky. Can an investor reconcile "moral grandstanding" with better financial performance when adopting ESG standards? While recognizing that social outcomes have economic value, can they be easily priced? In fact, how important are social outcomes to investors, versus governments?

Collecting and monitoring even basic data on ESG is part of the controversy. In addition to the lack of clarity in defining terms, there is the pressure to broaden the scope of the evaluation criteria to include indirect social outcomes, monitor user responses, or include data that some would label as anecdotal and subjective. Two additional considerations affect measuring social outcomes. First, social outcomes can be influenced by reducing social damage activities, such as crime prevention measures, or improved health care. Second, the value of many social outcomes is not evident until well into the future. The economic value of a social initiative might be stated today, but the true value would be subject to some methodology to account for the time value of money between today and some future period (Preqin, 2021).

While the claim is that good governance is good business, it is not clear just how much governance is required to sustain a successful private business. Governance of most large public institutional investors is subject to external policies and procedures usually established by governments, and similarity with the publicly listed funds. Both are subject to disclosure requirements, auditing requirements, risk assessment, the influence of stakeholders or shareholders, active board of directors, and clear roles and responsibilities for those in senior management roles. Listed funds can also be subject to market pressures, swings in the market that they have no control over, and the relatively short tenure of many CEOs. Pension funds are usually held to high standards of governance, but this varies across countries and a review of pension funds, both public and private, might arrive at a conclusion that their core mission of creating shareholder value is clouded by other less transparent factors and external influences.

In September 2022, Larry Fink, CEO of BlackRock, the world's largest asset manager, and a strong proponent of ESG, issued a 10-page letter in response to 18 U.S. state attorneys general who sent a letter to BlackRock essentially arguing that its goal of moving toward a net-zero economy conflicts with its fiduciary duty (Primack, 2022). All 18 were in Republican-controlled states. Two states, Texas and West Virginia, also banned state entities from doing business with BlackRock, arguing (incorrectly) that Blackrock boycotts fossil fuel company investments. After disputing the "boycott" accusations, the firm wrote: "We believe investors and companies that take a forward-looking position with respect to climate risk and its implications for the energy transition will generate better long-term financial outcomes."

ESG and institutional investment in infrastructure

It is important to distinguish between ESG, impact investing, and socially responsible investing (SRI), all three of which refer to forms of value-based investing (Fidelity Investments, 2022). ESG focuses on companies making an active effort to either limit their negative social impact, deliver benefits to society, or both. Impact investing draws a direct connection between value-based priorities and the use of investor capital with some ability to quantify positive societal benefits. Socially responsible investing (SRI) involves screening out businesses that conflict with an investor's value choices. Examples would be the "sin stocks" that generate profit through the sale of tobacco, alcohol, or weapons. SRI is the simplest and often least expensive approach to value-based investing.

It is reasonable to assume that the growth in institutional investment in infrastructure would be paralleled by a growth in the application of all three value-based considerations, but with clear emphasis on ESG. This growth is driven by three factors: the inclusion of ESG factors across a portfolio of investments; the characteristics of infrastructure as an asset class; and the realization by investors that they have a responsibility to society driven by increasing public and private pressures to address environmental concerns and practice responsible investing. There are also regulatory and legislative requirements driving this initiative. For investors, ESG offers two potential advantages that can enhance shareholder value by delivering better risk-adjusted returns.

The first is enhanced risk mitigation by favoring investments that attempt to minimize negative effects on society and emphasize positive effects. This imbeds ESG in the risk management processes of an organization. Negative factors to be considered can range from worker strikes, the possibility of litigation, to unfavorable media coverage and publicity. The second advantage is the opportunity to respond to the increasing pressures for responsible investing coming from investor groups including pension funds, endowment funds, and public interest groups. Today, the drive towards ESG comes from a collection of pressures from institutions, asset managers, and the industry.

Infrastructure as an asset class has unique characteristics that make ESG a useful tool for managing certain kinds of risk. The relevant characteristics include the long life span of infrastructure assets, the central role that infrastructure plays in economies and the well-being of society, and its impact on the environment and communities. ESG has the potential to address risk factors, financial and non-financial, across the life cycle of an infrastructure asset that can include political, regulatory, and reputational risks to which investors with long-term investment horizons are exposed. Table 17.1 is an example of what can be included in an ESG risk matrix.

There are two ESG issues for institutional investors in infrastructure.

- First is the impact of ESG factors on financial performance which is central to the debate on sustainable investments (OECD, 2020). Some studies suggest that ESG factors tend to reduce risk and generate higher returns, but lack of data and consistency of criteria hinders this research. A recent study by EDHEC addresses the relationship between ESG and the market value of infrastructure investments (EDHEC, 2021). This is a key question that institutional investors and prudential regulators need answered in order to integrate ESG into their financial decision-making processes.
- A second issue deals with the framework and tools for ESG analysis. Several international standards and tools have been developed in order to integrate sustainability and resilience aspects into infrastructure development and support ESG infrastructure asset analysis. Among the most used are the Sustainable Development Goals (SDGs), the IFC Performance Standards, Sustainable Accounting Standards Board (SASB), the International Organization for Standardization (ISO), the International Framework for

Table 17.1 Example of ESG Risk Analysis for Infrastructure

ENVIRONMENT	SOCIAL	GOVERNANCE
• GHG emissions • Air quality • Energy management • Water and wastewater management • Ecological impacts • Impacts of climate change	• Health and safety • Employee engagement, diversity, and inclusion • Human rights and community relations • Customer privacy • Access and affordability • Product quality and safety	• Materials sourcing and efficiency • Business ethics • Supply chain management • Competitive behavior • Critical incident risk management

Source: Schroder AIDA. Policy Risk & Impact Assessment Report: Infrastructure Finance. 2019.

Integrated Reporting (IR), the Global Reporting Initiative (GRI), and the UN-supported Principles for Responsible Investing (PRI). However, there is a preference among investors for internally developed methodologies and this complicates access to reliable information and limits evidence-based research on ESG topics. Other impediments to comparative performance measurements for ESG include the difference in disclosure requirements across types of funds, who is responsible for gathering data within a fund, how well they track exposure and performance, what tools they use, and what benchmarks they utilize.

In recent years, a number of tools have been developed to support infrastructure investors and developers in the implementation and monitoring of ESG factors in investment analysis, with the aim of producing a rating, certification, or financial impact assessment. Evaluation tools can be useful during the due diligence process, for benchmarking investments or projects, as a tool for reporting and stewardship, and for considering how a project addresses various ESG criteria across a portfolio. Among the major issues is lack of a common definition and a common set of metrics for measuring exposure to ESG risks in infrastructure; the heterogeneity of the infrastructure landscape; the lack of quality data and information required to perform analyses; the ability to quantify and monetize ESG criteria; transparency in valuation methodologies across the industry; investors' understanding and confidence in ESG valuation; and the costs of ESG analysis.

A central role for ESG

Despite the drawbacks, criticisms, inconsistencies, and uncertainties that accompany the use of ESG by infrastructure investors today, ESG continues to play a central role in delivering sustainable infrastructure and its importance will only increase in the future. ESG will be an essential part of gaining community support and establishing a social license to operate (SLO) for new infrastructure and investing in adapting existing infrastructure to meet climate-related goals. ESG may still be in the very early stages of evolution, but it has already garnered strong support from public and private players in the infrastructure.

Among the many challenges that must still be addressed are the lack of common definitions and metrics for measurement, the lack of quality data and information, the ability to quantify ESG criteria in financial terms, lack of transparency, particularly within the unlisted funds, and the ability to build confidence in and understanding of ESG valuation methods. The work of both the Task Force on Climate-related Financial Disclosures and the Coalition for Climate Resilient Investments, previously discussed, should address some of the shortcomings. There is also a lack of clarity as to the role of government and the ESG conditions that governments may apply to infrastructure projects in which they are involved, or that take place in its territory. Some of these challenges will not be easy to address given that infrastructure investments are often undertaken through private market transactions in which proponents want to protect confidentiality and limit disclosure of what they see as competitive information.

What ESG represents to the infrastructure industry is an opportunity for sharing knowledge and best practices that could promote improved public–private collaboration. It can also provide a seat at the table for industry representatives when it comes to shaping policy frameworks, government regulations, and procurement practices. The good news is that there are an increasing number of organizations and regulatory frameworks aligning with a common objective to develop and refine metrics to improve the efficacy of ESG in the investment world.

Part V
Investing in new infrastructure

18 Greenfield projects

The infrastructure gap will not be closed without a very significant increase in investments in greenfield infrastructure in the decades ahead. Nor will the infrastructure requirements in the emerging economies be addressed and climate change mandates met if greenfield infrastructure is not ramped up. But the prognosis is not good. The COVID crisis has severely limited the investment capacity of most governments to ramp up spending, however, the funding problems preceded the pandemic. For the past seven years, private investment in new infrastructure has remained stagnant and lower than it was 10 years ago. The $156 billion invested in infrastructure projects by private investors in 2020 represents 0.2 percent of global GDP, far shy of the 5 percent of global GDP (combining public and private investment) some studies indicate is required to close the infrastructure gap (GIHub, 2021). The GIHub notes that, in 2020, private sector investments in infrastructure in emerging economies declined by 28 percent, continuing a pre-pandemic trend.

Unfortunately, most of the institutional capital invested in acquiring infrastructure assets is directed at secondary stage assets, with a small amount in brownfield projects. Brownfield refers to existing assets or structures that require capital improvements, repairs, or expansion, most of which will be directed at operational enhancements. Secondary stage refers to fully operational assets with no immediate or significant investment needed.

The possibility of institutional investors being involved in greenfield projects has been a long-running and recurring topic of discussion, with no real evidence of progress. It is not for want of trying. Across the world, governments, multilateral financial institutions, banks, and institutional investors have over the past 10 years launched multiple initiatives to attract private capital into greenfield infrastructure, using both equity and debt (Lavanchy, 2020). Despite many attempts, greenfield projects remain the least favored preference among investors across the infrastructure choices available to them. For example, greenfield deals in Europe involving private investors fell from 43 percent of all infrastructure deals in 2010 to 30 percent in 2019, compared to investment in operational assets with no capital expenditure requirements—secondary stage deals—which more than doubled, rising from 287 to 668 (GIHub, 2021).

DOI: 10.1201/9781003396949-24

The number of pure greenfield deals only rose from 252 to 291. The common refrain from investors is that the greenfield opportunities that are presented to the market are not bankable (Figure 18.1).

What does "not bankable" mean? It has multiple dimensions and will mean different things to different investors. It can refer to prevailing market conditions where there are more opportunities elsewhere for financial and contractual innovations that can mobilize private capital for large-scale, capital-intensive projects with a strong revenue potential. It may be the result of counterparties being in poor financial shape and having a high credit risk with no de-risking mechanisms in place. There may be issues that can delay a project moving into an operational stage, or that can compromise steady and predictable returns that investors require. A project may also lack subsidies and support mechanisms such as enhanced payment security measures, guaranteed off-take agreement, lump sum termination payments, or compensation for undue delays in receiving clearances and permits. Even the absence of a pipeline of bankable greenfield projects can be a significant deterrent to private investors. Most of these impediments require risk mitigation measures as part of project preparation activities to move a project toward bankability.

Challenges for the private sector with greenfield infrastructure

The massive pressure for new infrastructure stems from several interrelated factors: population growth and migration flow into cities combined with deterioration and obsolescence of existing assets. Globalization affects infrastructure, giving rise to new societal needs and preferences, promoting the

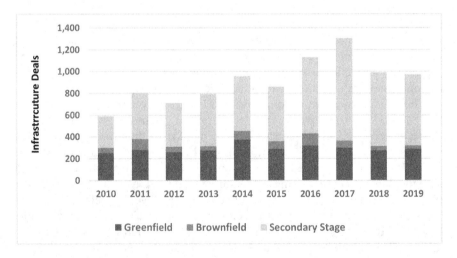

Figure 18.1 Infrastructure deals by project stage in Europe, 2010–2019.

Source: Preqin 2020 Global Infrastructure Report.

movement of goods and services across nations, increasing consumption of resources, and often manifesting in geopolitical posturing. New infrastructure projects are getting larger, more complex and costly, more capital intensive, and now expected to meet new social equity and public welfare requirements. New infrastructure, whether social or economic, is essential to the growth and prosperity of communities, states, and countries. Pressure is also mounting to rehabilitate and upgrade existing infrastructure to meet climate change mandates and strengthen the resilience of cities.

Most of these needs must be met with greenfield projects, some carried out exclusively by government, others by the private sector, and some in partnership arrangements. What all of this activity implies is access to huge amounts of capital and extraordinary levels of technical expertise, very little of which currently resides within the public sector.

Greenfield needs will not be met unless there are significant changes in how governments and the private sector approach these challenges. Investors in infrastructure have enjoyed almost two decades of low-maintenance investments, predictable risk profiles, and strong and consistent returns that have weathered chaotic periods (Brinkman and Sarma, 2022). Infrastructure has consistently outperformed other asset classes in a portfolio context. But conditions that have existed over the decades are rapidly changing, and not necessarily for the better. There is now evidence of growing political, economic, and social unrest across nations, introducing uncertainties in the market that are exacerbated by inflation and rising interest rates. Markets continue to suffer from disruptions caused by COVID and the implications of the war in Ukraine are not fully revealed, or even understood. Shifts in infrastructure markets in response to these pressures are indicative of structural shifts that will impose requirements for new ways of doing business for infrastructure investors from those that got investors this far in such short a time.

Significant changes for investors will be in the risk profiles in the asset classes themselves. Risks will proliferate and intensify and likely not be matched by enhanced returns. Uncertainties will reshape the investment horizon and begin to turn the attention of investors to greenfield opportunities in several sectors. The future can no longer rely on an almost exclusive focus on secondary transactions. There is even the possibility that new risks are taking shape that could lead to value destruction in what were once considered low-risk core assets. An example is the impact of climate change and what is currently happening across the globe in terms of hurricanes, flooding, droughts, forest fires, temperature extremes, and rising oceans.

In contrast to existing infrastructure that may be threatened by these tragic climate-related events, greenfield infrastructure presents new opportunities driven by new technologies in energy transition, resiliency, sustainability, electric mobility, and digitalization (Brinkman and Sarma, 2022). But these opportunities also require taking on new risks and managing higher levels of risk than in the past.

For investors, meeting the challenges and capturing the opportunities of greenfield infrastructure imply not only changes in business practices, but undertaking a cycle of learning with its successes and failures, developing new global networks and partnerships, changing risk/reward expectations, and exploring new markets not previously considered. Next-generation markets already on the horizon include hydrogen capture and distribution, rail technology, electric vehicles, and 5G telecommunications networks. Next generation may also imply larger and more complex infrastructure projects to flood-proof coastal cities, move entire settlements inland, address the impacts of drought on agricultural production, and make settlements more self-sustaining and detached from global networks and regional utility grids. Achieving net-zero targets in the next two decades cannot be achieved without expensive and extensive investments in new infrastructure.

It may be difficult to incorporate greenfield projects into the same organization or division that handles brownfield or secondary acquisitions. The professional skills sets will be different, as will the risks and the risk/return profiles. Venturing into greenfield may require setting up a separate division, establishing new fund structures, or building separate investment platforms through which to source and develop greenfield opportunities. Much of this will require collaboration with other partners including other investment funds, operating firms with unique technical knowledge and skills, and government entities.

Public–private partnerships

Where do PPPs fit into the discussion as they are a preferred procurement model by some governments for greenfield infrastructure? PPPs have allowed governments to spend more on infrastructure than they could otherwise afford, and have two appealing features (Engel et al, 2014). First, most of the investment in PPPs is not included in public debt, nor does it contribute to the fiscal deficit, at least in the short run. Second, PPPs are normally not subject to the procurement oversight and other budgetary controls that apply to other forms of procurement. These advantages apply to creating the asset, but largely ignore the fiscal impact of PPPs once the project is delivered. Other reasons for governments to choose PPPs include lack of internal resources and capabilities, inability to manage certain risks, and long-term management of the asset considering a government's consistent record of deferred maintenance. In some developing economies, PPPs are heralded as a means to avoid corruption, particularly where there is international involvement and oversight of the process.

PPPs have played an important role in delivering much need social infrastructure such as hospitals, schools, and public facilities in Europe, the UK, Australia, and Canada, and economic infrastructure in Latin America and Asia. They have been used in developing countries to build toll roads, with India being a good example. To date, PPPs have made limited inroads in the

U.S. relative to the size of their infrastructure market. However, interest is growing, and the new infrastructure legislation recently enacted by the U.S. Congress, plus recent successes such as the PPP contract for the multi-million-dollar redevelopment of LaGuardia Airport, Terminal B, and the most recent approval of a PPP for a new \$4.2 billion terminal at New York City's John F. Kennedy International Airport may serve as catalysts for expanding the PPP market in the U.S.[1]

Unfortunately, a recent court case in Pennsylvania struck down a Macquarie-led PPP to repair and replace nine bridges across the state using tolling as the source of revenue[2]. The court ruled against the state's Department of Transportation, the sponsor, and in support of a group of local opposing communities who claimed that the correct consultation process was circumvented by the PPP procurement model. Politics was at play. Shortly after the court ruling, the state passed new legislation that requires bridges to be repaired without charging motorists tolls, as well as ensuring the state's general assembly has more time to assess future transportation projects.

Devising a capital structure for PPPs

In promoting PPPs, it is in the interest of government to seek the lowest cost solution, one in which they are satisfied that they have achieved the lowest possible cost in utilizing private capital, while achieving the desired level of risk transfer to a private partner. Government wants to lock down and secure cost certainty for the life of any arrangement. They base their forecast on the assumption that little will change over time, that they have alleviated themselves of certain risks, and that the asset will perform to the same level 30 years hence when it is returned to public ownership when the concession agreement expires. In contrast, a private partner seeks to maximize the return on their investment, be adequately compensated for the risks they acquire, and have an opportunity to drive up returns with some value-added opportunities, or increased revenues throughout the term of the arrangement. Governments are most comfortable with availability-based PPPs that offer more certainty over revenue-based models that incorporate more variables including revenue volatility. In comparison, the private sector prefers revenue-based arrangements whereby they have an opportunity to enhance returns by driving up use of the asset or driving down operational costs. It is challenging to align the respective interests of the public and private partner in such an arrangement, but that is what PPPs hoped to achieve.

In undertaking a PPP, government will usually devise a capital structure that represents a combination of equity and debt, both by the public and the private partner across the time span of the project. Figure 18.2 depicts the complexities of PPPs with respect to both inflows and outflows of capital. The largest need for capital is during the construction period, usually with a combination of equity and debt. Repayment occurs at substantial completion, which certifies the end of the construction period. Revenue flows should

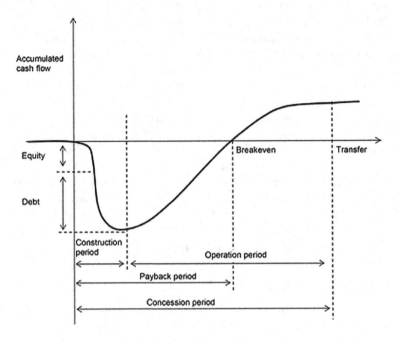

Figure 18.2 Typical cash flow profile for a PPP.

Source: Zhang, 2009. Win–Win Concession Period Determination Methodology. Journal of Construction and Engineering Management 135(6): 550–558.

commence upon substantial completion, or shortly thereafter, whether availability payments or user-generated cashflow. There can be a period following substantial completion and the beginning of cashflows where there is insufficient revenue to cover costs and this may require some form of bridge financing, a very costly form of debt.

This flow of cash is referred to as the "cashflow waterfall", and depicts revenues, debt obligations, and operating costs, in a pecking order that places all contractual obligations above equity[3]. The equity bucket is always last to be filled, and the debt buckets are first to be filled. Once all obligations have been met, including debt payments and payments to bond holders, operating costs, taxes, and fees, any remaining funds will be distributed to the equity holders as dividends. This process is governed by a complex network of legal agreements and underwriting requirements that will include insurance and performance ratios including debt service coverage ratios (DSCRs), all of which will focus upon preserving the rights and protection of debt holders. Risk mitigation may also include an early warning system that will provide debt holders with sufficient time to address problems and even trigger step-in provisions for debt holders if financial distress or failure is a possibility. Risk mitigation can include a debt service revenue fund, prefunding of expected major maintenance and repair, and mechanisms for dispute resolution and well-defined procedures in case of default, termination of one of the parties, restructuring, or a sale or transfer of ownership.

The capital structure will reflect what is happening in the capital markets at the start of a project, the health of national and local banks, the interest level of the private sector, the balance sheet of a government and its credit rating, political considerations, legal and institutional constraints, the degree of competition among bidders, and the track record with PPPs. Through this mix, government must arrive at a total project cost acceptable to both sides.

Capital allocation decisions cannot be made independent of expected cash flows, and a projected cash flow waterfall that can fulfill obligations to repay contracted debt and equity-generated expectations of the private partner. Capital structure decisions are not only financial decisions, but also risk management decisions. If the cash flow waterfall is endangered, or fails, the private partner may walk from the project and forfeit their equity, an option that government does not have. Government is the risk taker of last resort in a PPP, a factor that is seldom priced in their penchant to drive down costs and minimize private capital.

Living up to expectations has been a challenge for many PPPs. Several high-profile PFI projects in the UK have been subject to renegotiation or bail out by government, fueling strong negative sentiments. There have also been refinancing windfalls in the operational stages of a PFI by the private partner, reinforcing suspicion of a private partner's ultimate intentions in the deal. PFIs have come under fire from the British taxpayer for their perceived inflexibility, the Government being seen to "mortgage" future generations whilst at the same time generating considerable surplus profits for the private sector investors in the up cycle. There are also situations where the private partner encountered unforeseen and costly construction problems, or projects stalled or failed because of the Great Financial Crisis of 2008. There continues to be polarization surrounding the use of the PPP procurement model and these divisions may be part of the reason why PPP investments in the last decade are now a small fraction of what they were at their peak in the early 2000s[4]. Skeptics of PPPs include politicians, public officials, some of the large construction firms, as well as taxpayer groups. Politicians and public officials dwell upon the value for money issue, a perennial debate prompted by the claim that the costs of capital raised in the public sector will always be cheaper than private capital, assuming that this public capital actually exists.

Some of the large global contracting firms now realize that bids on large complex projects have seriously mispriced construction risk, with no recourse to negotiation in the agreements. SNC-Lavalin (SNC), one of Canada's largest infrastructure contractors, announced in 2019 that it was withdrawing from bidding on a number of high-profile public–private partnership projects in Canada. This move by SNC reinforces Fitch Ratings' view that the market for PPPs has become increasingly competitive and that project completion is one of the most significant risks to a project[5]. There are three multi-billion PPP public transit projects in Canada, two of which are well behind schedule (Toronto, Ontario and Edmonton, Alberta), by years and not months, with no completion date in sight for either. Each appears to be seriously over budget and what were deemed partnerships are now likely headed to litigation. The third PPP, in Ottawa, Ontario, was recently completed but became the subject of a public enquiry due to the many problems that continue to plague the project[6]. All three represent a summary of what can go wrong with a PPP.

In the U.S., one of the largest global markets for infrastructure, PPPs have not gained a foothold. This can be attributed to the highly decentralized political structure of the country. This political structure embodies the strong independence of state governments, federal and state tax codes, and the wide variation in political and social attitudes that exist across states. Each U.S. state retains a strong level of autonomy, particularly in government procurement and contractual practices. Involvement by the federal government in PPPs is negligible. In fact, what are often referred to as PPPs in the U.S. are often standard design-build contracts that are publicly financed. Equally challenging is the high level of hesitation in the U.S. about private sector management of critical public services. This resistance extends to almost any type of governmental function where there is a perception that the public interest is being compromised and corporate profits will be achieved by diminishing the quality of service. There is also a prevalent fear that private sector involvement could result in a drastic reduction in the number of government employees. Nor can the influence of powerful lobby groups, including unions, be overlooked.

Unique to the U.S. market is the absence of pressure to turn to PPPs to access private capital. Local governments and states can access capital through bond issues that are exempt from both federal and state taxes, and therefore can finance projects at lower interest rates than generally available to the private sector. Through what are referred to as "Muni Bonds", state and local governments have a direct line to Wall Street to raise public funds.

Outstanding municipal issuances of Munis totaled $3.7 trillion in 2018, down from a peak of $4.1 trillion in 2010 (Lucas and Montecinos, 2021). Annual issuances have fluctuated around $400 billion in recent years, with revenue bonds comprising over half of that total and general obligation bonds accounting for most of the rest. Bonds may be issued on behalf of a private partner and when they are tax exempt, they are called Qualified Private Activity Bonds. In addition to conferring a tax advantage, being granted access to Muni financing signals implicit government credit support that further lowers the interest rate. This additional advantage can be quantified and considered a subsidy.

When it comes to balancing the interests of the private and public sectors in any "partnership" arrangement such as a PPP, there is a wide variation of opinion. In fact, PPPs are not partnerships in the legal sense, rather they are a complex set of legal agreements assigning risks and rewards between participating parties that cover every aspect from project development and delivery through operations and governance and extending over a long period of time. There is also the question of whether these legal agreements can "remain whole" over 30- or 35-year concession agreements[7]. However, it is not the technical or legal complexities of PPPs that are the real challenge. The challenge is dealing with the accusations of "illegitimacy" of fixed arrangements through which governments burden future generations with long-term contracts with the private sector that are devoid of transparency and governance structures that can protect the public interest (Hodge, Greve and Boardman, 2010).

Concerns with the "illegitimacy" of PPPs are significant in the emerging economies where procurement laws may lack strength and recourse to an institutional framework to uphold the law. In developing countries, where there is general agreement on the need to attract and utilize the private sector and their capital sources in new infrastructure projects, many of these economies have little choice other than to use a PPP, particularly when they must rely upon a multi-national bank or other international provider for financial and technical assistance.

In terms of their contribution to investments in physical and social infrastructure, estimates indicate that PPPs represent approximately 10 percent of the yearly investment in infrastructure by developing countries' governments and approximately 3 percent of global infrastructure spending (Fabre and Straub, 2021). It is also estimated that two-thirds of all public investment was channeled through state-owned enterprises (Deep, 2022). The majority of new public infrastructure is still financed by national, state, and local governments with direct allocations from public budgets and awarded through traditional forms of procurement such as engineering/procurement/construction (EPC) contracts.

Compiling accurate figures on world infrastructure spending and how much is private versus publicly financed is challenging. In lieu of real data, there are estimates prepared by a few global consulting firms. From the available data, it is estimated that PPP spending accounts for about 3 percent of global infrastructure spending, and 8 percent of private infrastructure spending (Engel et al., 2014). According to Airoldi et al., global public and private spending, excluding telecoms, averaged about $2.7 trillion in 2008–2010, broken down into transportation ($1,040 billion), social infrastructure ($490 billion), water and waste ($160 billion), oil and gas transmission ($190 billion), and electricity ($810 billion) (Engel et al., 2014). Global capex spending on telecommunications is estimated at about $300 billion in 2011. Therefore, yearly global spending on infrastructure is estimated at about $3 trillion, or around 5 percent of world GDP.

According to Airoldi, private spending, excluding corporate financing, represents about one-third of total infrastructure spending. Private infrastructure is funded through PPP project finance, through other forms of project finance, and by corporate financing, numbers for which are impossible to compile as they are imbedded in corporate balance sheets. Corporate financing of infrastructure is omitted from Airoldi's analysis, but one must assume it is a large number. According to George Inderst, total project finance for infrastructure projects around the world was about $382 billion in 2012, and total project finance for infrastructure projects varies between $280 billion and $320 billion (Inderst, 2013). According to Inderst, PPPs represent between $60 billion and $110 billion per year of project finance. About 75 percent of PPP spending is in the transportation sector and another 20 percent in financing infrastructure to deliver government services.

In the European Union, infrastructure PPPs emerged in the 1990s and grew until the GFC in 2008. Since then, numbers have declined from a high of 129 projects in 2007, worth $26.8Bn, to just 39 projects in 2018 worth $15.76Bn (Inderst, 2013). Since the 1990s, 1,841 PPP projects were undertaken in the entire European Union, valued at $411.2Bn, a very small fraction of total EU investments in infrastructure during this time. As a benchmark, in the period 1995–2014, the annual investment in just roads in the European Economic Association (EEA) averaged $119.65Bn annually[8]. In developing countries, the Private Infrastructure Advisory Facility (PPIAF) keeps a database of PPP projects classified by type of investment[9]. In the period 1990–2018 there were a total of 1,760 PPP projects with a total investment of $534.7Bn, half of which was invested in roads.

Challenges for the public sector in greenfield infrastructure

The future of greenfield projects in both developed and developing countries will rely on four things:

- Evolution of new investment vehicles to accommodate an increasing and active role for the private sector.
- Involvement of the private sector from the time a project is conceived to ensure that projects launched by governments are "bankable".
- Acknowledgment that investors will continue to view investments in infrastructure as a business proposition in which capital appreciation, along with income, drives up returns.
- Finally, recognition that the success of any greenfield venture will ultimately rely on people whether these are partners, advisors, technical support personnel, service providers, contractors, or the management and leadership team. People make the difference in achieving ultimate success.

The most significant challenge facing governments that recognize the need for private sector involvement in addressing their infrastructure needs is that of aligning interests. From a policy framework downward, there must be an alignment on the optimal level of private-sector participation and risk transfer throughout the life cycle of the project (Beckers et al., 2013). Too often governments portray the role of the investor as an instrument to address budget constraints, close a funding gap, and transfer risks that they are not capable of handling.

The topic is seldom approached as partnering with the private sector to deliver a cost-efficient project in a timely manner that meets customer needs and expectations. Too often, key decisions on design, development, and operation have already been made before an investor is introduced to the project. This approach falls far short of establishing an optimal level of private-sector participation that can meet requirements on both sides. It also illustrates the vast differences in how the public and private sectors view risk. An investment partner faces major consequences when a project encounter cost overruns or construction delays. Consequences for the investor get translated into yield compression, liquidity problems, lower credit ratings, and reputational damage. Construction and commercial risks can have massive financial consequences for investors, resulting in project termination or even bankruptcy.

The public sector must realize that private investors don't just accept risk; they manage risk through a complex system of enterprise risk management that embodies the culture of the organization. They expect to be amply rewarded for taking on this responsibility and don't see the cost of capital as the only benchmark against which compensation should be measured. The public sector must also acknowledge that there are risks they cannot transfer, or where risk transfer would be very costly. In this case, risk mitigation may involve de-risking strategies that can range from guarantees on revenue thresholds to accepting responsibility to resolve construction-related disputes that can affect costs and timing. The best position for the public partner is to assume that risky situations are generally easier and less expensive to prevent than they are to solve (Beckers and Stegeman, 2021).

An area where the performance of the public sector in promoting greenfield solutions has been woefully inadequate is in the application of design innovation. This is where the early involvement of investors pays dividends, particularly where these investors represent prior relevant experience or application of a particular technology. An empirical study in the UK in the hospital sector found that PFI schemes had limited success in encouraging private investors to put forward innovative infrastructure designs (Gil and Beckman, 2009). The study revealed that the terms of the procurement process and of the long-term management contracts discouraged promoters from innovating and taking risks in design. Even when new ideas were considered, they incurred the risk of being eliminated in value engineering exercises when project budgets were tightened.

Incentivizing private capital in greenfield infrastructure

There is an overall desire in most parts of the world to tap into the vast pools of private monies, notably pension funds, insurance companies, and equity funds, to address the infrastructure gap, but with limited success. Private investors in infrastructure primarily deal in the stock of existing infrastructure assets—through secondary transactions—whereas closing the infrastructure gap is a matter of increasing the flow of new infrastructure projects (greenfield projects).

In developed countries, most of this new infrastructure must be funded from the public purse using conventional procurement models. Squeezing in private money will not be easy. Developing countries neither have access to the magnitude of public funds they require, nor the years of experience with traditional procurement models they can rely upon. They are dependent on PPPs and the involvement of international advisors and private partners that can bring money and expertise to the table.

The REM project, Montreal, Quebec, Canada

An example of a new frontier in greenfield development is the Réseau express métropolitan (REM) project in Montreal, Quebec. In June 2015, the Québec National Assembly passed Bill 38 "An Act to allow the Caisse de dépôt et placement du Québec (CDPQ) to carry out infrastructure projects"[10].

CDPQ thereafter created a new subsidiary, CDPQ Infra, dedicated to the development of greenfield infrastructure projects. The bill granted powers to CDPQ Infra to enable the execution, management, and funding of major public infrastructure projects, in partnership with the Caisse. One of the biggest projects for CDPQ Infra is the Réseau express métropolitain (REM) in the City of Montreal, Quebec. This is a revenue-based PPP with the Province of Quebec underwriting a significant portion of the revenue risk. CDPQ Infra is designing, financing, building, and operating a fully automated metro line 67 km in length with 26 stations and connecting downtown Montreal, the South Shore, the North Crown, the West Island, and the international airport at Dorval. It was budgeted at CAN$6.3 Bn with the financial structure summarized in Table 18.1.

It is too early to comment on the success of the REM project as it is still in construction. However, it represents an adaptation of the PPP model and addresses risk mitigation by assigning all functions to a single private entity,

Table 18.1 Financial Structure of Réseau Express Métropolitain Project

Sources of Financing CAN$	
$2.95Bn	**CDPQ Infra**—Equity 70% of equity
$1.28Bn	**Government of Quebec**—Equity 30% of equity
$1.28Bn	**Canada Infrastructure Bank** (CIB) Secured loan
$295Mn	**Hydro Quebec** Electrification of transport—Commercial agreement with a major client
$512Mn	ARTM payment to replace future revenue that CDPQ Infra would have received as a result of the increase in land values
$6.3Bn	**REM Construction Costs**

involving two levels of government in de-risking: the Quebec government underwriting a portion of revenue risk and collaboration with the Canada Infrastructure Bank on a guaranteed loan with a below-market interest rate. CDPQ has already approached other countries with this version of PPP model and sees it as one solution to enhancing the use of PPPs to deliver public services in a cost-effective manner.

CDPQ's foray into other greenfield projects involves building a greenfield platform centered on its acquisition of the North American business of Plenary, a leading international investor, developer, and operator of public infrastructure.[11] By acquiring Plenary Group (Canada) and Plenary Americas, CDPQ acquired an operating business as well as a controlling stake in Plenary's existing public–private partnerships comprised of 36 projects. This also provides the basis for partnering with Plenary in other countries. CDPQ maintained its 20 percent ownership interest in Plenary Asia Pacific.

Notes

1 Sourced at www.reuters.com/world/us/faa-oks-new-42-billion-terminal-new-yorks-jfk-airport-2022-11-17/#:~:text=New%20York%20state%20Governor%20Kathy,be%20used%20by%20JetBlue%20(JBLU
2 Sourced at www.infrastructureinvestor.com/court-strikes-down-macquarie-led-ppp-in-pennsylvania/
3 See "Features of a Cash Flow Waterfall in Project Finance. Mazars." https://financialmodelling.mazars.com/resources/features-of-a-cash-flow-waterfall-in-project-finance/
4 Public support for the Public Finance Initiative (PFI) in the UK never recovered after the bailout of the London Underground PFI in 2008, which cost taxpayers somewhere between £170 million and £410 million according to the National Audit Office (NAO) in the UK. The Guardian newspaper has published several articles highly critical of PPPs (The Guardian, 2018).
5 Sourced at www.fitchratings.com/research/infrastructure-project-finance/snc-pulling-out-of-ppps-symptom-of-completion-risk-challenges-06-08-2019
6 See www.ottawalrtpublicinquiry.ca/documents/final-report/
7 To "remain whole" infers that a single entity (i.e., PPP comprised of an extensive collection of intertwined legal agreements covering all aspects of the entity), can remain in force as a single unit over the span of the agreement.
8 The European Economic Association (EEA) is a professional academic body which links European economists. It was founded in the mid-1980s. The EEA links the EU member states and three EFTA states (Iceland, Liechtenstein, and Norway).
9 Public-Private Infrastructure Advisory Facility (PPIAF) is a global technical assistance facility managed by the World Bank on behalf of donors. Established in 1999, it aims to help eliminate poverty and achieve sustainable development in developing countries by facilitating private sector involvement in infrastructure.
10 CDPQ is Canada's second largest public pension fund after the Canada Pension Plan and is very active in both real estate and infrastructure
11 Leo Kolivakis. CDPQ Acquires Plenary Americas. Pension Plus. March 12, 2020. http://pensionpulse.blogspot.com/2020/03/cdpq-acquires-plenary-americas.html

19 Challenges in project delivery

Construction remains one of the most stagnant industries across the world and in all sectors from housing to building dams and power plants. The industry relies on processes that often date back centuries (McKinsey, 2017). Throughout the world, the delivery of large engineering projects is plagued by such challenges as lack of productivity, an aging and shrinking work force, dated management practices, an aversion to adapting new technology, and an attitude that each project is a one-off prototype. A McKinsey study in 2017 estimated that, globally, the construction sector labor-productivity growth averaged just 1 percent, compared to 2.8 percent for the total world economy and 3.6 percent for manufacturing (McKinsey, 2017) (Figure 19.1).

The lack of productivity in construction caught the attention of the New York Times in an article by Ezra Klein and based on a research paper released by the U.S. Bureau of Economic Research (Goolsbee and Syverson., 2022). The authors of the research paper in their conclusion state that, "Measured productivity performance in the construction sector has been unusually awful for 50 years.... The sector's ability to transform intermediates into finished products has deteriorated.... The productivity struggle is not just a figment of the data. It is real".

Global challenges in large-scale project delivery

The processes by which infrastructure is delivered are the opposite to that of automobiles or electronics where the entire process is carried out largely under a single team reporting to a single owner, end-to-end. Automobiles and electronics involve a complex network of suppliers and sub-contractors, spread across the world, but industries in these sectors have slashed the time and cost of project delivery by adopting sophisticated logistics chains and refining how they deliver product. For example, car makers have accelerated new-model development by more than 50 percent (McKinsey & Company, 2017). Leading manufacturers have halved lead times and doubled productivity and output by introducing automation, machine learning, and smart technologies, as well as introducing strong partnership networks, greater agility, and flexibility,

DOI: 10.1201/9781003396949-25

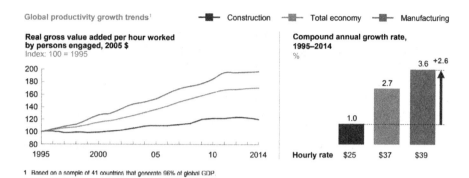

Figure 19.1 Global productivity growth trends.

Real gross value added per hour worked by person engaged, 2005 $
Index: 100 = 1995.

Source: Exhibit E1 from "Reinventing Construction: A Route to Higher Productivity", February 2017, McKinsey & Company, www.mckinsey.com. Copyright (c) 2022 McKinsey & Company. All rights reserved. Reprinted by permission.

and launching new manufacturing processes. The best that construction has achieved is the introduction of material handling equipment, handheld battery-powered tools, prefabricated form work, and limited use of robotics. Three-dimensional printing of components may be on the horizon but still a long way off. Some cost savings have been achieved by moving the assembly and finishing of components from in situ to off-site factories and the use of pre-assembled and value-added components. It is too early to assess the benefits of what is referred to as PropTech that currently applies almost exclusively to the real estate industry[1].

Project delivery for infrastructure remains a highly fragmented and transaction-driven process that can be broken down into four parts, each with the potential for significant value gain or loss as a project proceeds through the various stages. Impediments to progress in project delivery are not easy to address and this is not exclusive to emerging markets. Construction processes are deeply imbedded in institutional frameworks, building regulations, local labor practices, and cultural norms. Each project starts from scratch, often seen as one-of-a-kind with limited benefit from previous project experience, largely void of reliance on shared knowledge, and with an implementation process that is highly fragmented. Design is usually separated from procurement and construction, except in the case of design-build. Project delivery involves a myriad of parties, both public and private, many of whom will not have previously worked together and may never work together again.

Unfortunately, the construction industry worldwide is prone to widespread corruption given the large sums of money involved and the lack of transparency in many transactions. Corruption seriously distorts pricing.

Construction is notorious for the high level of off-market activities and unrecorded cash transactions. For many contractors and suppliers there are low barriers to entry and lax or opaque controls over their businesses. These are not criticisms levelled only at emerging economies—they apply to the most advanced economies. Information sources on corruption include Transparency Internationals' Corruption Perception Index[2] and JLL LaSalle's Global Real Estate Transparency Index Alternatives and Infrastructure[3]. These indices can be correlated with the Global Competitiveness Report[4] issued annually by the World Economic Form.

Project delivery also depends on multiple and complex legal agreements that cannot anticipate conditions that will inevitably arise almost daily during construction. These agreements are only as good as written and enforceable by the institutional framework that lies behind each one. This framework must support processes for conflict resolution, arbitration, payment claims, and legal challenges. It must be capable of offering resolution in a reasonably short time, and not impose unnecessary delays on a project. There is also the need for government representatives to act as responsible owners and managers, exercising their decision-making as independently as possible from their political masters.

These challenges are exacerbated in risk transfer agreements that can prompt confrontation and time-consuming dispute resolution processes when problems arise. This is what makes the Alliance model an appealing alternative in some projects whereby risks are shared. Introduced in Australia, and now widely used for large infrastructure projects in that country, Alliance contracting is a form of relationship contracting characterized by a culture of collaboration and cooperation between the parties delivering a project (Green and Chua, 2018). The parties to an Alliance are normally the purchaser of services (the owner) and one or more service providers or non-owner participants (NOPs) such as head contractor and operator. The parties' interests are aligned, and risks are shared through incentives offered by the owner for how well the project is delivered, as measured against agreed objectives.

Project delivery is a challenge across the globe, resulting in significant cost overruns and delivery delays for large infrastructure projects. McKinsey found that nine out of ten infrastructure megaprojects faced cost overruns that added, on average, 70 percent to the original budget (Garemo et al., 2015). They also found that 61 percent of these projects exceeded the original schedule.

Project delivery challenges unique to the emerging economies

Getting past the first hurdle, that of solving the financing conundrum for a project in an emerging market country, opens the door to a new host of challenges in getting something built. Risks can originate with the lack of competition among contractors capable of handling large-scale projects, a poorly trained workforce, lack of experience in managing large-scale projects among the various parties including architects, engineers, and consultants, not having access to dependable suppliers and sub-trades, and an underdeveloped institutional support framework. In addition to the potential for corruption, this translates into a set of risk factors that are difficult to anticipate and even harder to price. These risk factors are multiplied where projects involve the refurbishing of existing brownfield assets.

A research paper published in Public and Municipal Finance documented some of the challenges faced by the Independent Development Trust (IDT) in infrastructure delivery of the provincial government of KwaZulu-Natal (KZN), South Africa (Khumalo et al., 2017). The authors surveyed a group of participants comprised of project managers, quantity surveyors, engineers, architects, and project managers working with IDT. The nature of the research was quantitative and data analysis used descriptive and a bit of inferential statistics to arrive at some generalizations and conclusions. The study was able to affirm that there are major inefficiencies in the current infrastructure delivery model of the South African government. Major causes identified include factors such as delays in payments, poor planning, and an absence of professional ethics and standards exercised by professionals involved. The study concluded that "Clearly the infrastructure delivery model requires a new trajectory in tackling the under-development and triple challenges of poverty, unemployment and slow economic growth".

Another research paper examined the relationship between construction delays and cost overruns in Rwanda, more specifically in the Regional Cybercrime Center (Amanya and Njenga, 2022). To quote the authors:

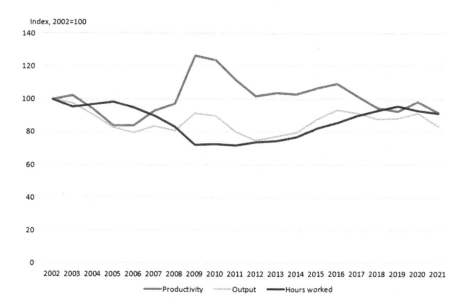

Figure 19.2 Productivity for highways, street, and bridge construction in the U.S. (2002–2021).

In spite of the interventions by Government of Rwanda in private sector for supporting construction projects, most areas in Rwanda are marred by projects that are delayed, disrupted, and uncompleted. Explanations for the stages of each building project vary and are sometimes attributed to poor financial muscle, lack of human and material resources with which to execute the projects. During the period 2012-2015, 65.7 percent of public construction projects in just one district of Rwanda were delayed. Another example is the Kigali Convention Center, originally scheduled to open in 2011, but was postponed until 2016. The construction of Bugesera International Airport is yet another illustration, it was supposed to be completed by the year 2016, nonetheless has yet to break ground (Ndundo & Mbabazi, 2016). A construction project analysis was carried out to review the primary reasons for delays in Rwanda. In accordance with statistical analysis of 15 commercial building projects completed between 2010 and 2020, none of them met planned project schedule. It was discovered that 15% of these projects were canceled, and 75% required more time to be completed.

Impact of project delivery on sustainability requirements

Large institutional investors today are likely to subject their potential investments in infrastructure to an ESG filter. On the environmental side, this will require construction firms to adjust practices to achieve climate-related targets and in doing so will require expert analytical and technical skills (McKinsey, 2022). This becomes even more complicated when the construction industry is required to achieve net zero targets, perhaps as early as 2030 in some countries. Developed countries are tackling this mandate by going after the low-hanging fruit and avoiding the more complex problem of overhauling the entire process by which infrastructure projects are designed and built. Unfortunately, there is scant evidence that the construction industry has the will to change, and the manifestations of this posturing will be more pronounced in the emerging economies, not because they have a stronger will to resist change, but because they lack the knowhow and skills to do so. An example would be carbon reduction, a highly complex undertaking requiring the assistance of highly trained experts.

Social infrastructure will have to operate in a world 1.5–2.0°C warmer and with wetter climates. Even behavioral changes among users must be anticipated in response to climate change. Sustainability will require a more open mind to keeping and improving upon what we have and adapting existing assets to meet new needs either with new technologies or modified requirements. Any changes from current practices will require targets, timelines, data collection, and means of measuring progress.

New approaches to address sustainability will require collaboration between the public sector and private investors to de-risk investments, the establishment of long-term partnerships among like-minded participants, an open mind to considering alternative solutions, and the introduction of new technologies such as carbon capture and hydrogen cells. A significant hurdle in emerging economies will be addressing human resource needs to address an impending shortage of skilled labor and available expertise in all facets of project delivery.

Notes

1 Property technology (also known as proptech, PropTech, prop-tech, and real estate technology) is used to refer to the application of information technology and platform economics to the real estate industry. https://en.wikipedia.org/wiki/Property_technology
2 www.transparency.org/en/cpi/2021
3 www.joneslanglasalle.com.cn/en/trends-and-insights/research/global-real-estate-transparency-index/greti-global-rankings-and-methodology
4 www.weforum.org/reports/the-global-competitiveness-report-2020

20 Investing in emerging and frontier economies

The term "emerging markets" (EMs) generally applies to both emerging markets and frontier economies, although wide differences exist across countries. The term, emerging market, was coined by Antoine W. van Agtmael in 1981, a member of the World Bank's International Finance Corporation (IFC) (Athukorala et al., 2007). Emerging market economies are classified in different ways, by different observers[1]. Levels of income, quality of the financial system, or growth rates are all popular criteria. The exact list of emerging market economies can vary depending on who you source. The International Monetary Fund (IMF) classifies 23 countries as emerging markets, while Morgan Stanley Capital International (MSCI) classifies 24 countries as emerging markets. There are some differences between the two lists. Standard and Poor's (S&P) classifies 23 countries and FTSE Russell classifies 19 countries as emerging markets, while Dow Jones classifies 22 countries as emerging markets[2]. Collectively, these countries include large and small economies, with growth rates that range from low to high.

The emerging markets (EMs) today are quite different from what they were just a decade ago (Manulife, 2021). Commodity-oriented, extractive industries have given way to the rapid ascendancy of companies in technology and consumer-oriented fields. This shift is accompanied by expanding wealth in the EMs and the emergence of EM-based tech companies that have established a global market share and, in some cases, have become close rivals to developed-market peers. The EMs today are increasingly recognized as the engines of key transformational changes that are reshaping global investment markets. These markets are aware of the infrastructure needs driving transformational changes and increasingly large institutional investors in infrastructure are taking positions in several EM countries. However, recent global events have introduced volatility in performance within the EMs, driven by changes in monetary policy, increasing political uncertainty, and deteriorating conditions for international trade (Melas, 2019). External financial flows to EMs were stagnating even before the COVID crisis of 2020–2021 (Inderst, 2021). The pandemic has only exacerbated weaknesses in these markets and the impacts of the Russian invasion of Ukraine are yet to be determined.

DOI: 10.1201/9781003396949-26

The evolution of the emerging markets

Despite the huge need for investment in the EMs, development spending on infrastructure comes predominantly from public budgets. In the late 1980s, a few pioneering investors ventured beyond developed equity markets and sought opportunities in Latin America and Southeast Asia. It was not an easy move due to the considerable obstacles they faced throughout the investment process, from difficulties in obtaining information, access to trusted local advisors and partners, rampant inflation, operational challenges, and how to repatriate the proceeds of their investments (Inderst, 2021). The potential for corruption was always a significant risk. The lure was evidence of high economic growth that could translate into high earnings growth, attractive valuations, and the possibility of superior portfolio returns over time.

The MSCI Emerging Markets Index was created in December 1987 to address the need of investors for investment research and performance benchmarks in selective emerging markets. At its inception the index covered 10 markets (Argentina, Brazil, Chile, Mexico, Portugal, Greece, Jordan, Malaysia, the Philippines, and Thailand) and represented less than 1% of the global equities' universe. A great deal of political and economic turmoil followed closely on the heels of this launch, including the demise of the Soviet regime, the return to democracy in much of Eastern Europe, the fall of apartheid in South Africa, and the beginnings of China's return as a competitor on the global stage; all contributing to the rapid expansion of the emerging markets universe. By 1992, just 5 years after its launch, the MSCI Emerging Markets Index had expanded to cover 13 countries and represent 5.3 percent of the global equities' universe. In the next 5 years the index further expanded to include 28 markets, representing 6.8 percent of the global equities' universe. But rapid growth in some of the EMs came with a high price, including the Asian financial crisis of 1997, the Russian debt default of 1998, the burst of the high-tech bubble in 2000, and the bear market in the early 2000s. By the turn of the century, the EMs lost much of their appeal, and by 2002 they only represented approximately 4 percent of the global equities universe. There has been a turnaround in the last decade with increasing market accessibility and the liberalization of the domestic Chinese market that had a dominant effect on the emergent market segment. Chinese equities now make up 20 percent of the emerging markets index with a forecast of this rising even higher (Inderst, 2021) (Figure 20.1).

Investment needs

The need for new infrastructure and the refurbishment of existing infrastructure is very high in the developing world and represents a significant growth constraint. Lack of modern infrastructure limits the ability to deal with the pressures of urbanization and meet crucial development goals, and relegates environmental concerns to a relatively low priority. According to Bhattacharya

Figure 20.1 The emerging markets sub-groupings of ODA eligible countries.

Source: OECD Development Co-operation Report 2014: Mobilizing Resources for Sustainable Development (ODA refers to Official Development Assistance).

et al. (2012), while it is inherently difficult to make precise estimates, in part because of the gaps in data, investment spending in infrastructure (excluding operation and maintenance) in developing countries would need to increase from approximately $0.8–0.9 trillion per year in 2012, to approximately $1.8–2.3 trillion per year by 2020, or from around 3% of GDP to 6–8% of GDP[3]. This includes about $200–300 billion to ensure the infrastructure lowers emissions and is more resilient to climate change. In aggregate, the incremental investment spending across the EMs is estimated at around $1 trillion a year more than what was spent in 2012 (Bhattacharya et al., 2012). The top priorities, and representing the bulk of spending needs, are electricity, water (upstream and downstream), and transportation.

In 2012 it was estimated that East Asia would require the majority of this investment (35–50%) with some 85 percent of this amount required by low- and lower-income countries. Approximately 45–60% of the investment requirement would be in the electricity sector, including generation, transmission, and distribution networks. The remainder would be split relatively equally between the transport, telecoms, and water sectors.

Meeting most of these needs will largely depend on greenfield projects initiated and financed by governments and relying upon domestic budgets to cover costs. Many governments will not have the resource capacity to front-end these projects and financing possesses its own set of challenges. In addition to normal commercial and physical risks, greenfield infrastructure projects require large risk capital for upfront investment associated with the preparation, development, and construction phase. Additionally, many projects face risks around revenue streams associated with policy uncertainties and affordability (e.g., user fees) making many projects unbankable. In addition to capital costs, there will be a significant impact on domestic budgets for ongoing maintenance and operating costs. These costs could double annual spending requirements (Rozenberg and Fay, 2019). Sustainability requirements to meet low emissions and higher efficiency improvements, plus addressing a broader set of social benefits including resiliency, biodiversity, security, safety, and public health, could add another 10–15 percent to the capital cost (Bhattacharya et al., 2012) (Figure 20.2).

Of the estimated $0.8–0.9 trillion per year invested in 2008, the majority ($500–600 billion) was financed by domestic government budgets, 20–30 percent (approx. $150–250 billion) by the private sector, an estimated 5–8 percent (approx. $40–60 billion) through developed country ODA and MDB financing, and perhaps 3 percent (less than $20 billion) from other developing country governments (Bhattacharya et al., 2012). What is clear is that the existing institutional framework available for financing infrastructure in the EMs is not capable of meeting the scale of investment required on two counts. First are the conservative limits on how much debt a country can take on to finance their infrastructure and investment growth. The World Bank-IMF

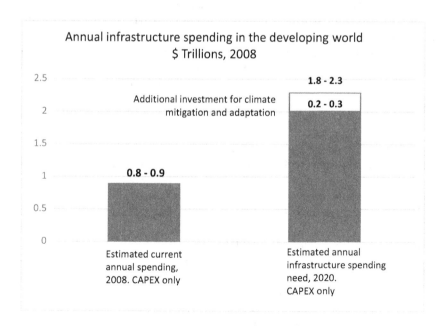

Figure 20.2 Infrastructure spending in the developed world.

Source: Bhattacharya et al. (2012). Infrastructure for development: Meeting the challenge.

Debt Sustainability Framework provides a guideline that the present value of a country's external debt should not exceed 30–50 percent of its GDP, and that debt servicing should not exceed 25–35 percent of government revenues (IMF-World Bank, 2023). Many EMs are already at this threshold. Second, this framework seldom provides the right mix and scale of finance to deliver the infrastructure investment required. This mix is affected by the mix of public and private financing, the risks involved, uncertainty of costs and revenues, perceived upside potential, and the various stages of a project from inception to completion.

A major impediment for most EMs is access to sufficient capital to develop and prioritize a pipeline of projects in addition to project preparation of specific projects. As previously noted, project preparation can be cumbersome, time consuming and costly, prone to using approaches that are excessively averse to risk, and with a lack of understanding on how to "crowd-in" private capital and structure a bankable deal. There are also public policy and political issues that can complicate the situation.

Investing in infrastructure for the EMs is not just about addressing infrastructure needs. It also is about achieving goals, having a clear policy framework in place, and improving the efficiency and effectiveness with which a country can pursue its goals. It is estimated that new infrastructure could cost low- and middle-income countries anywhere between 2 percent and 8 percent of gross domestic product (GDP) per year depending on the quality and quantity of service aimed for and the spending efficiency achieved to reach that goal (Rozenberg and Fay, 2019). But investing in infrastructure is not enough; maintaining it also matters. Ensuring a steady flow of resources for operations and maintenance is a necessary condition for success. Good maintenance also generates substantial savings and can reduce the total life-cycle cost of transport and water and sanitation infrastructure by more than 50 percent.

Role of governments

The role of governments in closing the infrastructure gap in the emerging economies is a complex topic well beyond the scope of this book. This is a topic that encompasses a group of countries that are not homogeneous, represent a broad array of social, economic, and political challenges, and for which there is wide disparity in the stages of development of their infrastructure across the spectrum.

However, it is worthwhile to consider ways to assist governments in emerging markets as far as increasing the opportunities for private sector financing for infrastructure to address their most pressing infrastructure needs. In addressing these needs three things are clear:

- First, governments have the primary role in launching new infrastructure projects and, if private financing is required, then projects must be structured to appeal to private investors as an investible asset class.

- Second, they must convince their traditional partners in the development finance institutions, such as the MDBs, to approach the market with a portfolio or pipeline of projects attractive to the investment markets.
- Third, they must address the governance and capability gaps that too often hinder private-sector investment. Governments must also be cognizant of the risks that private investors will want to avoid. A de-risking mechanism will be a necessary tool to mitigate significant risks during the construction and delivery stages of greenfield projects.

An important role for governments in the emerging economies is the promotion of infrastructure that addresses public health and welfare concerns, environmental considerations, and climate change risks as complementary to achieving economic prosperity or middle-class aspirations. This is not about spending more, but rather spending wisely. Private investors in infrastructure take a long-term view and want assurances that investments can meet ESG requirements, that infrastructure is being built in pursuit of goals in addition to growth, and that investments that promote social equity, environmental preservation, and even personal enrichment contribute to the long-term profitability of their assets.

Risks in the emerging economies

The cost of finance is a key driver of project feasibility, whether in developed or developing economies. Each project may involve a mix of government financing allocated from annual budgets, project loans, private sector financing from one or more investors, and international public finance. Each of these participants will have a suite of products at their disposal and a clear idea of the risk/return profile that each will consider. Even in the case of governments, lenders such as commercial banks or a bond market will want to know what risks the debt entails, whether the debt meets their underwriting standards, and the impact of this debt on the balance sheet of the country. As the financing gets layered, each participant will negotiate for a preferred position in the payment schedule and attempt to match the level of financial need with an expected return.

An underlying assumption in this risk identification exercise is that most infrastructure projects in the emerging economies are "greenfield" projects which have a much higher risk profile than "brownfield" projects, particularly at the front end. It is also likely that many of these projects will be structured as public–private partnerships. Many private investors will have limited experience with greenfield projects and, particularly with PPPs, this translates into a demand for much higher returns on their capital to compensate for the risk involved. Many private investors will take a position that it is prudent to take a pass on investing in the EMs, given the alternatives.

The greater the risk transferred to a financier, the greater the expected return on any invested capital. The risk-return profile of projects will change substantially, both according to the nature of the project and according to the phase the project is in. For each of these combinations, financiers will have to provide different types of finance to match the risk-return profiles as well as cash flow. Higher risks will lead to higher costs of financing, particularly if commercial finance is required.

Risks fall into three categories: (1) project preparation; (2) delivery of new infrastructure; and (3) operation and maintenance (Bhattacharya et al., 2012). Figure 20.3 provides an illustration of the risks and financing decisions at each stage of a project's life cycle.

In addition to the traditional risks that most investors would be familiar with, there are five risks that private investors will prioritize that are unique to the EMs and present a significant barrier to private investors. It is difficult to price the risk for any of the five.

- **Sovereign risk:** the threat of a government repossessing an asset (nationalization) without due compensation. This can happen with a change of government or in response to intense political pressures on a national government. The partially built international airport in Mexico City is an example (Foster, 2021).
- **Regulatory risk:** a government changes the rules or regulations despite contracts in place. Examples would be a new government coming to power and imposing new taxes, reducing tariff rates, preventing a sale or transfer, or restricting the withdrawal of equity.

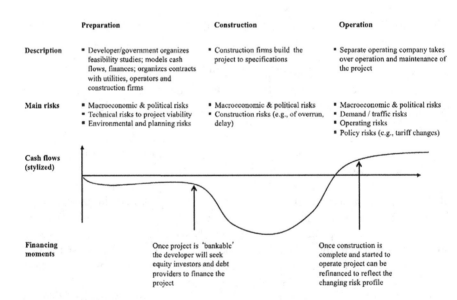

Figure 20.3 Risk and financing considerations at different phases of a typical infrastructure lifecycle.

Source: Bhattacharya et al. (2012).

- **Reputational risk**: where corruption plays a significant role. Odebrecht in Brazil is a well-known case of widespread corruption in the construction of new infrastructure projects in South America (Gallas, 2019).
- **Resource risk:** the availability of qualified domestic partners, availability of local consultants, and a skilled domestic workforce.
- **Liquidity risk:** the ability to liquidate the asset at some future point in time. An example is the ability to repatriating profits from offshore investments.

The impact of risks is not uniform across the various stages of a project and, depending on each stage, a risk may diminish in importance and another risk may take on increasing importance. Risks gets balanced as a project proceeds and it is the overall risk-return profile that matters to an investor. The challenge for investors in any emerging market is that the risks in almost all the above categories are quite high in comparison to developed countries, and benchmarks seldom exist to serve as comparators for pricing these risks. To compensate for these higher risks, investors and financiers will demand high returns on their capital or stay away altogether.

Notes

1 www.investopedia.com/terms/e/emergingmarketeconomy.asp
2 www.investopedia.com/terms/e/emergingmarketeconomy.asp
3 Estimates adapted from Rozenberg and Fay (2019). Beyond the Gap: How Countries Can afford the Infrastructure They Need while Protecting the Planet. World Bank Group.

21 Financing infrastructure in developing countries

There is common agreement that ways must be found to attract and retain private capital to fund infrastructure in emerging economies. This is capital to supplement the scarcity of public funds and to promote sustainable infrastructure that can address climate change and other threats to lives and livelihoods in socially and economically deprived communities. However, there are significant challenges to moving the discussion beyond a general commitment to achieving a goal, and a paucity of evidence on progress to date upon which to bolster confidence.

Several clarifications are necessary as a prelude to discussing this topic. This this is where the distinction between financing versus funding infrastructure is important. As previously stated, these are not interchangeable terms. Infrastructure transactions in emerging economies are based on a mixture of financing sources and confusion arises in referring to public versus private financing. Public financing refers to a government contribution, either debt or equity, from monies in a capital budget or raised through taxes, fees, or other public sources of revenue. Private financing refers to all other sources of capital, most of which involves MDB, and multinational and bilateral providers.

Greenfield projects are the primary recipients of this capital. Projects may be revenue-based as in the case of a toll road or availability-based where there are no revenues such as a school. Most private investors, including the large institutional investors and the unlisted funds, have limited experience with greenfield projects, seldom see themselves in a lender role, nor are they attracted to bond-like payment structures such as availability payments. Institutional investors are primarily large-scale equity investors seeking opportunities to match their long-term liabilities against a revenue stream, and preferably with value-added opportunities. It is a revenue stream that they expect to manage over time to achieve a risk-adjusted return that meets their expected rate.

The key players in financing greenfield infrastructure in developing countries are the multinational development banks (MDBs) and the development finance institutions (DFIs). Both the MDBs and the DFIs have been instrumental in getting out the message that the mobilization of private investment capital is a key missing ingredient to addressing long-term infrastructure needs.

DOI: 10.1201/9781003396949-27

Unfortunately, the flow of this private capital over the past decade is negligible relative to what is needed, particularly in the low-income countries and to meet sustainability goals. George Inderst estimates the total flow of private capital to the emerging economies at approximately 0.2 percent of GDP annually (Inderst, 2021). Where private capital has been invested it is in brownfield projects, secondary deals, structured debt, and loan guarantees.

No one disputes the magnitude of need, nor is there a shortage of opinions on what might be done to attract private capital. In 2015, the heads of the African Development Bank, the Asian Development Bank, the European Bank for Reconstruction and Development, the European Investment Bank, the Inter-American Development Bank, the World Bank Group and the International Monetary Fund, issued a press release stating that achieving the Sustainable Development Goals (SDGs) will require moving from billions to trillions in resource flows (IMF, 2015). The press release notes that while there is no substitute for concessional resources, especially for the poorest, most fragile, or conflict-torn countries, marshalling other types of financing at the levels needed will demand greater efforts to unlock, leverage, and catalyze more public and private flows. Financing from private sources, including capital markets, institutional investors, and businesses, will be particularly important.

Mobilizing private capital

It is important to clarify definitions that pertain to any discussion on private capital in the emerging economies markets (Multinational Development Banks and Development Finance Institutions, 2021). It is also important to understand that discussions on private capital contributing to financing infrastructure in the emerging markets are built on a framework that measures the involvement of the MDBs and DFIs, both international and national, in facilitating private investments. The term private direct mobilization (PDM) refers to financing from a private entity on commercial terms with the active and direct involvement of an MDB leading to a commitment. Another term, private indirect mobilization (PIM) is financing from private entities provided in connection with a specific activity for which an MDB is providing financing, where no MDB is playing an active or direct role that leads to the commitment of the private entity's finance. PIM includes sponsor financing if the sponsor qualifies as a private entity.

The combined total of PDM and PIM defines the level of total private mobilization. Private co-financing (PCf) is an investment made by a private entity, which is a legal entity established for business purposes and financially and managerially autonomous from any national or local government. Public co-financing is the investment made by a public entity. Public entities include multilateral and bilateral financial institutions, export credit agencies, and any other institution whose primary purpose is to benefit or promote a specific national interest, regardless of ownership.

The total amount of the loan or equity being guaranteed by an MDB is counted as PDM. This includes both commercial risk guarantees and non-commercial risk guarantees. For commercial risk guarantees, the PDM is counted net of any amount reported as an MDB commitment. Guarantees by sponsors are counted as PIM and subject to attribution.

OECD uses a different approach to defining private capital mobilization to that of the MDBs, although the underlying principles are the same (OECD, 2022)[1]. The OECD approach does not differentiate between direct and indirect mobilization as with the MDB approach. The OECD attributes private mobilization to all official development finance interventions in a project; the MDB approach only attributes this amongst MDBs contributing to the joint report. The OECD approach is still evolving and currently their measures cover five instruments (guarantees, syndicated loans, shares in collective investment vehicles, credit lines, and direct investments in companies). The MDB approach covers all instruments. The OECD approach defines infrastructure to include only economic infrastructure (roads, energy, etc.), while the MDB approach includes some social infrastructure (hospitals, schools, etc.)

Whether using the MDB or the OECD approach to measurement, overall numbers on the levels of private mobilization are not encouraging, particularly when compared to levels of need. According to the MDB Task Force on Mobilization, in 2019, MDBs mobilized $63.6 billion of private finance in operations in middle- and low-income countries. While this amount represents an overall decline of 8 percent from 2018, it includes $6.7 billion mobilized for low-income countries, a significant increase of 21 percent. Mobilization in high-income countries also increased 22 percent from 2018 levels. Nearly half of the mobilization to LICs and MICs was generated by the World Bank Group (WBG). The breakdown by continents shows some $14 billion for Africa (0.4 percent of GDP), of which $6 billion is direct[2] (0.2 percent of GDP) (Inderst, 2021).

An OECD study in 2018 (OECD, 2018) presents a different set of numbers and this reflects the different scope, methodology, and terms and definitions. Over the period 2012–19, OECD estimates that $257.6 billion was mobilized from the private sector by official development finance interventions. The data show an upward trend between 2012 and 2018, followed by a 9 percent decrease in 2019 (Figure 21.1).

Blended finance

Blended finance is the use of catalytic capital from public or philanthropic sources to increase private sector investment in sustainable development. It is gaining increasing recognition as a catalyst, using public or philanthropic sources of capital to increase the flow of private capital into sustainable infrastructure projects in emerging economies. The two main investment barriers for investors that blended finance addresses are (1) high perceived and real

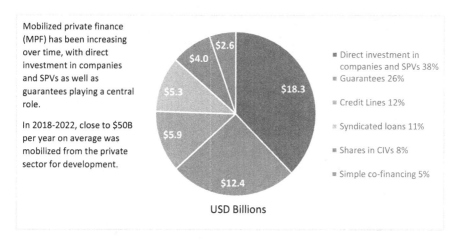

Mobilized private finance (MPF) has been increasing over time, with direct investment in companies and SPVs as well as guarantees playing a central role.

In 2018-2022, close to $50B per year on average was mobilized from the private sector for development.

- Direct investment in companies and SPVs 38%
- Guarantees 26%
- Credit Lines 12%
- Syndicated loans 11%
- Shares in CIVs 8%
- Simple co-financing 5%

$2.6
$4.0
$5.3
$18.3
$5.9
$12.4

USD Billions

Figure 21.1 Mobilizing private capital.

Source: OECD, 2022.

risks and (2) poor returns for the risk relative to comparable investment opportunities (Convergence, 2022).

According to Convergence (2022), to date, blended finance has mobilized approximately $172 billion in capital towards sustainable development in developing countries. Convergence identifies approximately 5,700 financial commitments to these blended finance transactions. Transactions range considerably in size, from a minimum of $110,000 to a maximum of $8 billion. The median blended finance transaction was around $64 million in total size (2010–2018).

Blended finance is a structuring approach (Figure 21.2), not an investment approach, and is distinct from impact investing which is an investment approach (Convergence, 2022). Concessional debt or equity has been the most common archetype, including first-loss debt or equity, investment-stage grants, and debt or equity that bears risk at below-market financial returns to mobilize private sector investment. There has been an increasing use of both concessional debt or equity and guarantees or risk insurance in recent years (Figure 21.3).

Three groups are active in blended finance: private investors with an explicit impact mandate (64 percent of total transactions), public investors with a development mandate (19 percent); and philanthropic investors (17 percent). The dominant public investors are the International Monetary Fund (IMF) and the Dutch Entrepreneurial Development Bank (FMO). The most active philanthropic investors in blended finance have included the Shell Foundation, Bill & Melinda Gates Foundation, Omidyar Network, and Oikocredit.

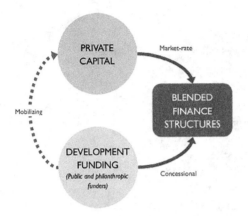

Figure 21.2 Blended finance is a structuring, not a financing approach.

Source: Convergence, 2022.

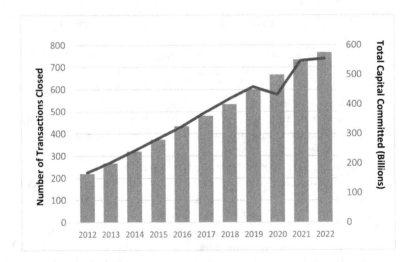

Figure 21.3 Growth in blended finance activities.

Source: Convergence, 2022.

As a structuring approach, Convergence identifies four common blended finance structures (Figure 21.4).

According to Convergence (2022), climate change has consistently been a thematic focus of the blended finance market since 2011, and climate-oriented transactions have accounted for 50 percent of deals launched annually (on

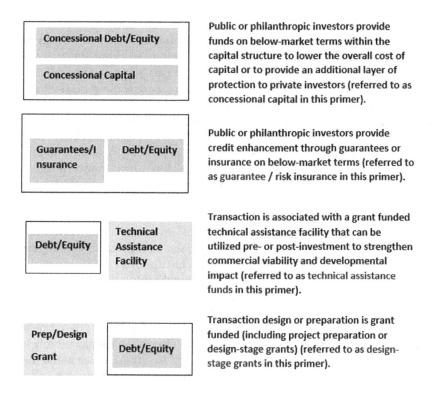

| Concessional Debt/Equity | | Public or philanthropic investors provide funds on below-market terms within the capital structure to lower the overall cost of capital or to provide an additional layer of protection to private investors (referred to as concessional capital in this primer). |
| Concessional Capital | | |

| Guarantees/Insurance | Debt/Equity | Public or philanthropic investors provide credit enhancement through guarantees or insurance on below-market terms (referred to as guarantee / risk insurance in this primer). |

| Debt/Equity | Technical Assistance Facility | Transaction is associated with a grant funded technical assistance facility that can be utilized pre- or post-investment to strengthen commercial viability and developmental impact (referred to as technical assistance funds in this primer). |

| Prep/Design Grant | Debt/Equity | Transaction design or preparation is grant funded (including project preparation or design-stage grants) (referred to as design-stage grants in this primer). |

Figure 21.4 Blended finance archetypes.

Source: Convergence, 2022.

average). These deals received over two-thirds of aggregate annual financing, notching an annual aggregate deal value of just under $7 billion. Concessional blended funds were used the most in lower-middle-income countries, and regionally in sub-Saharan Africa and Europe & Central Asia. Overall volumes of capital flows precipitated by blended finance are still very low. Sub-Saharan Africa has been the most frequently targeted region in blended finance transactions, with 45 percent of transactions. In recent years, Asia and Latin America have emerged as new frontiers for blended finance. Energy has been the most frequently targeted sector, followed by financial services.

Convergence (2022) indicates a decline in aggregate financing levels in the climate blended finance market in recent years—between 2019–2021, $14 billion was invested into climate blended finance transactions, compared to $36.5 billion between 2016–2018. The proportion of annual climate blended finance flows relative to the overall level of financing to the blended finance market declined from 74 percent between 2016–2018 to 61 percent between 2019–2021.

Investor potential in the emerging markets

Private investors remain reluctant to get involved in financing new infrastructure in the emerging economies. Funding opportunities, from the perspective of investors, are almost non-existent. For the past seven years and across the globe, private investment in infrastructure in the emerging markets has remained stagnant, and lower than it was 10 years ago (GIHub, 2021). About three-quarters of private investment in new infrastructure projects occurs in high-income countries and appeared to be unhindered by the pandemic. Half of this investment occurred in renewable energy generation. Middle- and low-income countries attract only a quarter of the global private investment in infrastructure projects and saw a 28 percent decline in private investment in 2020. Most of this investment occurred in the non-renewable energy and transport sectors, primarily in highways.

According to GIHub, around 80 percent of investment in infrastructure projects in primary markets is financed by debt, mainly from commercial banks, investment banks, and financial services institutions.

Projects in developed economies are increasingly using debt raised in the capital markets. Financing through green bonds has been rising in recent years, particularly in high-income countries. In middle- and low-income countries, non-private institutions like multilateral development banks (MDBs), export credit agencies (ECAs), governments, and others also play a significant role as financiers. In fact, 75 percent of private investment in infrastructure in middle- and low-income countries occurs in projects that involve both private sector and non-private sector financing.

Priorities for investment depend on the income group. In high-income countries, half of private investment in infrastructure projects occurs in renewable energy, while in middle- and low-income countries, two-thirds occurs in transport and non-renewable energy. In high-income countries, in 2020, almost 55 percent of the private investment in infrastructure projects went to renewable energy generation, and in middle- and low-income countries, that percentage was only around 20 percent. This compares to over 25 percent for non-renewable energy generation (Inderst, 2021).

Institutional investors are keen to find ways to direct capital into the emerging economies and this interest varies by country and sector. However, significant barriers remain that characterize the mismatch between investor objectives and development needs. Possible solutions may involve increased opportunities for direct investment; purchases in the secondary market of securities of large, listed companies particularly in financial, telecoms, and energy companies; involvement in private equity and credit transactions; use of impact funds; and structures using blended finance. The big question is one of timing. When will conditions be right, what opportunities may arise, and will business conditions be sufficiently attractive to private capital? These are questions that perhaps only governments can address.

Notes

1 OECD. Amounts mobilized by the private sector for development.
2 Direct mobilization is "financing from a private entity on commercial terms due to the active and direct involvement of a MDB leading to commitment." It consists of loans, equity, guarantees, risk and capital market products, and Islamic finance products (Inderst, 2021).

Part VI
New frontiers for investors

22 Non-tangible infrastructure assets

The assumption is always that infrastructure refers only to those assets that can be seen and touched—essentially physical things. However, the economy does not depend on tangible assets alone, even though infrastructure investments that national statistical offices measure have been, until very recently, all tangible assets. Recognition of the importance of non-tangible assets dates to the 1960s and 1970s when futurists such as Alvin Toffler began to talk about a "post-industrial" future and the so-called knowledge economy began to take root in the early 1990s with the emergence of the Internet in the early 1980s[1]. In 2006, Microsoft's balance sheet was around $250 billion, but its balance sheet indicated a valuation of around $70 billion, $60 billion of which was cash and various financial instruments (Haskel and Westlake, 2018). Plant and equipment were only valued at $3 billion, or 1 percent of market value. Why was Microsoft worth so much? The answer lies in its investments in research and development, product design, branding, supply chains, internal structure, and its human capital, all things that consumed time and money, but are not readily visible.

The OECD defines an intangible as "something which is not a physical asset or a financial asset, which is capable of being owned or controlled for use in commercial activities, and whose use or transfer would be compensated had it occurred in a transaction between independent parties in comparable circumstances" (OECD, 2014). In essence, the value of an intangible is dematerialized and becomes apparent only when, say, a company is sold and the market assigns a value to it. Specifically, national accounts compiled by statistical agencies usually include various types of IP, including patents, know-how and trade secrets, trademarks, trade names and brands, rights under contracts, and government licenses, as well as goodwill and ongoing concern value and investment in software. Corporations usually count as intangibles IP items including patents, copyrights, customer relationships, brands, and concession rights, as well as goodwill and investment in software. How they account for intangibles depends on a range of factors, including national tax codes, as well as accounting conventions and rules.

DOI: 10.1201/9781003396949-29

As Haskel and Westlake (2018) point out, "there is something fundamentally different about intangible investments, and that understanding the steady move to intangible investments helps us understand some of the key issues facing us today; innovation and growth, inequality, the role of management, and financial and policy reforms" (Haskel and Westlake, 2018). The authors argue that there are two big differences with intangible assets. First, most measurement conventions ignore them. Second, the basic economic properties of intangibles make an intangible-rich economy behave differently from a tangible-rich one. Mindset matters. Investing in intangibles is different from investing in tangibles. According to McKinsey, those who invest in intangibles "need to pivot to a mostly test-and-learn, agile culture" compared with 60 percent of low adapters (Hazan et al., 2021). Similarly, about 60 percent of top intangible adapters versus about 50 percent of low adapters agreed that it was important to "cultivate an environment to take risks, test and learn, and embrace failures" (Hazan et al., 2021).

The McKinsey article by Hazan attributes the differences to four things.

- Investments in intangibles tend to represent a sunk cost with a very limited resale market to recapture some of the investment. Patents might be sold, but most intangible investments are tailored to meeting specific needs of the business making the investment.
- Second, intangible investments have *spillovers*. There is a tendency for others to benefit from what were intended to be private and exclusive investments.
- Third, intangible assets are more likely to be *scalable* as exemplified by brand names, patents, and other forms of intellectual property.
- Finally, intangible investments tend to have *synergies* with one or another—they are more valuable together or in the right combinations.

Over the past 25 years, investment in intangible assets has risen steadily as a share of total investment in the United States and 10 European economies—Austria, Denmark, Finland, France, Germany, Italy, the Netherlands, Spain, Sweden, and the United Kingdom (Hazan et al., 2021). In 1995, the split was about 70:30 in favor of tangible investment; by 2019, the split was 60:40. It is notable that the intangibles' share continued to increase even in the face of major economic disruptions; indeed, some evidence indicates that that trend may have accelerated during the COVID-19 pandemic in 2020 and early 2021.

McKinsey identifies four major types of intangible assets (Hazan et al., 2021) (Figure 22.1):

- **Innovation capital.** Arising from investments that build a company's intellectual property (IP). This may include investments in R&D, new product development, process innovation, or new product interfaces.
- **Digital and analytics capital.** Arising from investments in building software such as the installation and maintenance of customer relationship

management software, developing databases, building a database or a data management platform.

- **Human and relational capital.** This spans two subcategories. First, organizational and managerial capital that includes investments that build individual or organizational skills through training, including advancing the skills of a workforce in a particular specialty. Second, capital associated with ecosystems and networks including activities related to developing and improving privileged relationships.
- **Brand capital.** Arises from investments in marketing and sales that build and improve brand equity.

The growth of intangible investments could radically change not only current definitions of infrastructure, but what kinds of infrastructure the economy needs to underpin future growth. The growth of investment in intangibles could also change the infrastructure debate and redefine the infrastructure gap. This gap largely refers to the need for physical things—bridges, airports, alternative energy sources, electrical grids, and telecommunications. Not that tangible infrastructure will become obsolete or result in more stranded assets. Much of this tangible infrastructure will continue to be important and make significant contributions to mitigating the effects of the rise in CO^2 and climate-related impacts. What intangibles signal is that the economy is changing and that technological advancements are gaining headway. The tangible infrastructures that will gain in importance are the investments that will make an intangible-rich economy more competitive. For example, telecom infrastructure will matter more in an intangible economy where more connections are required between people and businesses and transformations will be driven by new hardware and software, new work habits, and new standards and regulations. An example would be the extension of broadband into remote communities involving satellites or perhaps utilizing existing electrical grids in small towns and remote communities.

The emergence of data infrastructure

Over the past 25 years, the ICT sector, representing innovation-driven services, has delivered continuous innovation, and conceived new disruptive ideas. In the past 10 years, the sector has created and scaled successful platforms and markets that benefit from network effects that have achieved unprecedented scale at a magnitude never before observed. A prime example is the transmission and storage of data to support essential public needs, making these assets infrastructure-like by nature. Yet data infrastructure, which includes data centers, fiber-optic towers, and mobile phone antennae, has characteristics that set it apart from more traditional infrastructure assets such as power or transport (Aviva, 2020). This is creating opportunities for investors, although capturing them is not straightforward. Fast-changing technology and user demands, coupled with less regulation, make it difficult to assess long-term

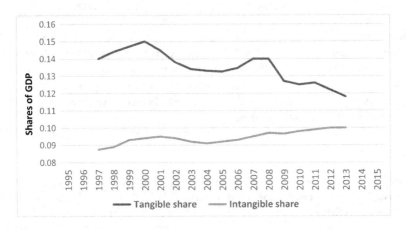

Figure 22.1 Tangible and intangible investments in Europe.

Source: Jonathan Haskel and Stain Westlake. 2018.

demand for specific assets. This explains why long-term investors are likely to focus on assets benefiting from contracted revenues or high barriers to entry, situated in a jurisdiction with clear regulation. However. data infrastructure falls short in terms of securing ESG credentials. Storing data requires vast amounts of energy. Data centers consumed an estimated 3 percent of the world's energy and could use one-fifth of all electricity in the world by 2025 (Aviva, 2020).

A recent example

Using existing railroad tracks and the same principle of "blocks" (only one train at a time is allowed in a given block), the recently unveiled Magnetic Railway Onboard Sensor (MAROS) uses information encoded in the very rails a train is running on in the form of something called their magnetic permeability (Economist, 2022). MARCOS identifies the train's exact position on the network, something that GPS or one of its equivalents cannot do due to tunnels and sets of tracks running in parallel. Through detector coils over each rail, and passing alternative currents through them, MARCOS picks up random fluctuations in the rail created at the time of production and this results in a unique fingerprint for a given block of rails. The system is expected to be on the market in 2025. By allowing trains to be run much closer together than is possible with current automatic control systems, it is estimated that MARCOS could increase capacity of a busy rail network by 20–30 percent. This is an example of using new technology to enhance the efficiency of a tangible infrastructure asset at a much lower cost than installing new control systems. MARCOS technology represents an investment in what would be considered non-tangible infrastructure.

Two challenges facing investors in intangible infrastructure

In the early years, even in the most advanced economies, investing in intangibles was something of a side show. However, the balance between tangibles and intangibles began to shift with a steady increase in the former, and by the mid-1990s investments in intangibles surpassed tangibles. This general trend is not to imply that the same held true for infrastructure. In fact, significant private investments in infrastructure did not appear until the new century and did not pick up pace until after the Great Financial Crisis of 2008–2009. Reasons for this growth can be attributed to the changing balance of services and manufacturing in the economy, globalization, and the increasing liberalization of markets, developments in IT and management technologies, and the changing input costs of services. What is difficult to determine is just what proportion of investment has gone to intangible versus tangible infrastructure over the past several decades. The assumption is that intangibles have become increasingly important in the infrastructure sector in driving the adaptation of new technology, particularly in the data transmission and telecom fields and in enhancing productivity in general.

Two complicating factors are going to give pause to any private investor venturing into intangible infrastructure (Haskel and Westlake, 2018). The first is the speed of technological change in the infrastructure itself as evidenced by the continued roll out of fiber-optic cable, 4G and 5G networks, and chip designs that enhance speed and storage capacity. Does an investor decide to invest now, or hold off until tomorrow or next year when prices may come down or new technology is expected to come to market? The second challenge is much tricker to deal with. New infrastructure relying on new technology is likely going to require new ways of working, in new locations, and with new processes. The introduction of electricity required radical shifts in how things were made to gain the full benefit of this new source of energy.

Opportunities to invest in intangible infrastructure will require flexibility, a willingness to take new risks on a "trial and error" basis, the introduction of new forms of measurement and assessment, and new management tools. It may also result in new ways of working, in different locations, and with related changes in human behavior that are difficult to predict. Intangibles will create new markets, scalable operations, and give rise to new synergies, new partnerships and collaborations, and different reward systems. These are not insurmountable challenges, nor are they going to slow the trend to increased investment in intangibles. Infrastructure investors are well positioned to capitalize on the many investment opportunities in intangible infrastructure that are already on the horizon, nor can they afford not to take the plunge given the magnitude of the capital they must invest in the coming years.

Note

1 January 1, 1983, is considered the official birthday of the Internet. Prior to this, the various computer networks did not have a standard way to communicate with each other. A new communications protocol was established called Transfer Control Protocol/Internetwork Protocol (TCP/IP).

23 The need for creativity and innovation

The competitiveness of nations has always relied upon continued improvements in infrastructure, but the pace of these improvements is never consistent across generations and always seems to lag demand. Public and private leaders make decisions that shape our infrastructure, the cumulative results allow cities to flourish, industries to thrive, economic competitiveness to improve, and social needs to be met. The first wave of innovation in infrastructure in recent times was tied to the Industrial Revolution and gave rise to canals, railroads, telegraph lines, coal-fired energy plants, and improved water and sanitation systems. These were physical systems that drove unparallel economic prosperity throughout the Western world, raised standards of living, and gave rise to a middle class located in cities. They also had unintended consequences, many of which remain with us today.

The second wave of innovation arose after WWII, much of it based on the rapid advancement of materials and technologies related to the War effort. This included new forms of fabrication, mass manufacturing systems, inventions such as radar and sonar, nuclear fusion, telecommunications, computing, jet aircraft, low-cost personal and commercial vehicles, and even the production of a type of housing that launched the post-War suburb in North America[1]. The suburb, perhaps the last great urban vision shaped by innovations of the time, was designed and executed in the middle of the 20th century. The suburb represented the confluence of automobile ownership by the masses, new highways, innovations in financing home mortgages, and production-line manufacturing of houses. The beginning of a third wave can be traced to the rise of a digital world in the 1980s and 1990s, and related global communication networks that eroded political boundaries and challenged the sovereignty of many nations. This third wave can be attributed to the digital era, combined with global trends. This is an era characterized by an aging population in Western nations, a shift from rural to urban areas across the globe, rising social and economic inequity, and the rapid ascendance of China and India in terms of their share of global GDP, population growth, and geopolitical influence.

A common trait across all innovations is that each one reduces the cost of something that was a necessity at that point in time. In doing so, innovations

DOI: 10.1201/9781003396949-30

can destroyed value but replace what it renders obsolete with new things of increased value, spawn new patterns of consumption, and introduce new forms of living and working (Verma, 2022).

People are changing, always have, and not always for the better. The concentration of wealth and the rise in inequality in many countries has stalled median wage growth from that of the postwar era. Nor has innovation and globalization delivered on its promises of a better life for all. Scarcity of critical resources including water, food, sanitation, education, and health care, combined with climate change, now require a complete rethinking of what infrastructure is required to move forward. We need infrastructure that can better serve cities, enhance urban resiliency in the face of natural disasters, deliver essential services, and address affordability issues for the disadvantaged. The good news is that the foundations for this new infrastructure are evident in the rise of knowledge-intensive industries, improved technical capacity within industries, investments in education, research, and development, and improved environmental sciences backed by extensive data collection. If there is a general theme to this third wave of innovation it is that we need infrastructure that will protect people and the natural environment for future generations (Tomer et al., 2021).

This does not imply abandoning or replacing infrastructure that has served us for decades. A priority must be adapting existing infrastructure to meet current needs and improve efficiency and productivity. This is where technology can be of immense benefit. Automation of transit lines can significantly increase rider capacity on existing lines and improve safety; monitoring of sewage can provide critical data on pathogen levels in a neighborhood; GPS can assist with monitoring traffic flows, improving mobility options, and support new payment systems for highway travel; hybrid systems of health care and education using the Internet can relieve pressure on existing healthcare facilities and introduce new forms of remote learning. The COVID-19 pandemic prompted an overnight shift to a digital world that would have otherwise taken years to accomplish. The pandemic also highlighted weaknesses in current infrastructure and demonstrated how quickly society can adapt to a digital world without undergoing radical changes in how we live and work.

Drivers of change

Brookings in their report, *Rebuild with Purpose* (Tomer et al., 2021), identify four cross-cutting forces that will directly impact infrastructure networks and operational systems for decades to come: climate resilience, digitalization, workforce, and fiscal health.

- **Climate resilience.** Chronic climate challenges—rising emissions, floods, and other persistent environmental risks—pose destructive threats to our future. The most visible manifestation is the frequency of storms and flood events, forest fires, and huge temperature swings. In the U.S., in the 1980s there were an average of 2.9 such disasters per year, with an average annual cost of $17.8 billion. By 2010, this rose to 11.9 such disasters each year with an average annual cost of $81.1 billion. Compare what happened in just one year, 2020, to the data on previous decades (Figure 23.1).

Incremental technological upgrades to address adaptation, for example, can include smart metering technologies, combined with predictive modeling and maintenance monitoring, and along with improved control and management by a utility responding to customer demands. These improvements can conserve water and energy, reduce costs, and more accurately identify future needs. Similarly, grid hardening adaptations such as burying wires, properly maintaining power lines, and installing sensors can identify and isolate dangerous threats, including wildfires.

Cybersecurity is now a dominant concern and crucial to protecting infrastructure systems from privacy concerns and political threats. Even materials can be adapted for better resilience to climate risks such as permeable pavement to not only reduce storm runoff, but to filter pollutants and allow stormwater to infiltrate back into local groundwater, often at a lower cost than conventional pavement systems.

- **Digitalization.** The diffusion of digital technologies in homes and businesses across the country affects what we do and how we do it. From computers and

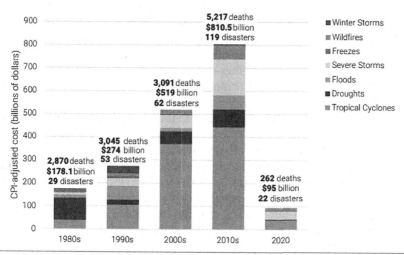

Source: Brookings analysis of NOAA National Centers for Environmental Information (NCEI) data
Note: Climate disasters refer to droughts, floods, freezes, winter storms, severe storms, tropical cyclones, and wildfires costing at least $1 billion each

Figure 23.1 Cost of climate disasters.

Source: Brookings Metro, Rebuild with purpose: An affirmative vision for 21st century American infrastructure.

smartphones to new transportation and manufacturing equipment, digital technologies have transformed the economy and everyday life, shifting how we communicate, travel, and buy and pay for things. Businesses have realized gains in productivity and efficiency as households have benefited from greater convenience and time savings. But digitalization has not improved outcomes for everyone, including those who lack access, affordability, or the skills to take full advantage of emerging technologies. Broadband connectivity gaps plague both households and businesses in both metropolitan and rural communities and limit opportunities to advance the digital economy.

Digitalization is radically changing retailing, logistics, financial services, and how and where we work. It also has the potential to improve productivity in the office and factory floor, as well as agriculture, all of which require improved infrastructure. Digitalization will allow some beneficial COVID responses to continue indefinitely in the post-COVID era, however, it is difficult at this point to predict which ones and their lasting influence since changes to behavioral patterns are involved. One of the greatest benefits and risks of the digital age is the abundance of data. New technologies continually emerge with the capability to collect constant, granular, and sometimes personal or sensitive information about people, places, and systems. This raises questions on cybersecurity and ethical practices.

- **Workforce.** Technology is one part of the innovation equation; the other critical ingredient is people. We lack the human capacity—a skilled infrastructure workforce—to design, construct, operate, and manage the infrastructure systems we have, never mind what we need to build in the future. This ranges from welders, electricians, plumbers, telecommunications experts, and equipment installers, to experienced project managers capable of overseeing multi-billion-dollar projects. Most labor markets are dominated by lower-skilled trades where pay schedules do not encourage long-term retention rates.

  Compounding the workforce problem is the rise of demand for both hard and soft skills and adaptation to new tools, technologies, and equipment that require digital literacy, ability to communicate and collaborate, and levels of emotional intelligence that can equip workers to move into leadership roles. The current workforce in many countries is aging, lacks diversity, and is not afforded opportunities for retraining and educational advancement. While governments espouse their commitments to improve infrastructure, seldom do they acknowledge that this cannot happen without people—a young generation attracted to high-quality, well-paying, and relevant work-based learning opportunities.

- **Fiscal health.** It is difficult to determine the veracity of the numbers used to dimension the infrastructure gap. The gap is predicated on *how* much we need to invest and *what* to invest in. It is light on justifying *why* and *where* to invest. The simple solution seems to be a call for spending more money, expanding capacity, fixing up what is broken, and addressing a policy framework that in many situations is outdated. This plea to spend more continues in the absence of money and fiscal hurdles that will not be easily overcome. More spending is necessary for repair and maintenance, but when it comes to new infrastructure, fiscal constraints will require doing more with less. This will require more effort to incorporate new technologies, business models, and emerging ideas. Addressing fiscal realities will require three things: a consistent and detailed approach to measuring infrastructure needs to target what is necessary, why, and when; modernizing existing infrastructure assets to better serve current user needs; and experimentation with new innovations and a recognition of the importance of human talent.

Expanding upon innovation

Innovation is not about a search for the proverbial "better mousetrap". Innovation in infrastructure must be formulated with sufficient flexibility to be robust and endure over long time periods, and be adaptable and scalable across a range of infrastructure sectors. Innovations must have synergies with existing infrastructure if they are going to work. Innovation is not just about hardware. Innovation can be broken down into three categories: policy innovation; financial innovation; and technical innovation (Cheong, 2015).

Financial innovation is largely a private-sector role, although governments play an important part through regulation of banking and financial services and the leverage they can apply through their credit ratings. An example being the development of a commercial infrastructure bond market by Infrastructure Ontario, an arms-length agency of the Ontario Government. This was created in response to the withdrawal of banks from infrastructure financing during the crash of 2008. This bond market has flourished since and funded subsequent PPPs in Ontario. Another financial innovation introduced by Infrastructure Ontario was a process by which the period between commercial close and financial close on a PPP proposal could be reduced to a matter of days or even hours. In many developed countries this transaction can take months for a variety of reasons. Financial innovations are particularly important in addressing climate-related infrastructure needs in the emerging markets.

Policy innovation is a key public sector role that embodies a commitment to more than just funding infrastructure projects. Policy innovation must be backed by a sound investment plan, tied to future user needs and drivers of economic growth. There must be a clear depiction of the respective roles of the public and the private sectors. Policy innovation should encourage experimentation and lay out a framework for working with governments.

Technical innovations may be somewhat easier to identify and describe once they emerge, but much more challenging when it comes to implementation. We see this with the advent of electrical vehicles and the struggle to install charging stations across a network of public throughfares and within buildings. Surprisingly, infrastructure is struggling to adapt new technologies to address shifting patterns of consumer use and expectations, and infrastructure systems remain among the least digitally transformed in the entire global economy (Smith, 2019). There is also the propensity with technical innovations to place undue focus on inputs versus tangible outcomes that include the user experience. This severely limits the value proposition for any new technology.

The infrastructure of tomorrow

The infrastructure of tomorrow will have a significant impact on what investors look at today. Investors realize that as much as 75 percent of the infrastructure we require globally over the next 30 years has yet to be built, or even imagined in certain sectors. They are also aware that ownership of much of this new infrastructure will end up in private hands and involve private investors working with partners, governments, or alone. Investors to date have largely focused on brownfield and secondary deals, much of which has been designed, built, operated, and maintained in the same way for decades despite huge technological and innovation gains. Much of this infrastructure is engineered for 50- to 100-year life spans and geared to the needs of the past and present users. Its longevity, a positive feature, also places it at risk in the face of constant and unpredictable change.

The infrastructure that has yet to be built and driven by new technological developments will begin to swing the pendulum for investors to seriously consider greenfield projects. The possibilities range from breakthroughs in artificial intelligence, electric vehicles, robotics, energy production and storage, and data-driven platforms. The downsides are the risks in introducing anything new and the ability of governments to update policies and regulations, recognizing that infrastructure investments often require close collaboration and support from governments. Just as infrastructure itself will be hindered by the human factor—lack of a skilled workforce—this will be even more severe among government departments and agencies involved in infrastructure, despite the best of intentions.

Private investors will be the beneficiaries of innovations if they get it right. Getting it right means betting on the right innovations, engaging users, figuring out potential revenue streams, and accounting for the new risks involved, while fully capitalizing on available opportunities. It also implies a much stricter discipline in addressing ESG requirements, particularly the "S" component, which means engaging with a much broader set of community stakeholders, Opportunities will increasingly involve a mix of investments in tangible and intangible infrastructure, whether digital networks, data centers, telecommunications, mobility, or energy. Capturing these opportunities will mean bringing large sums of money to the market and exploring new value-added propositions, in a more speculative decision-making framework. This will entail the involvement of new partnerships, minority stakes in start-up firms, acquiring new resources for an organization, and developing a more private equity-style approach to investment decisions. It will also entail a much more aggressive strategy to enter the emerging markets where new infrastructure will use innovation to leapfrog over infrastructure solutions that the West has relied upon for past centuries.

There are three sectors that have already captured the attention of infrastructure investors when it comes to innovation. The most obvious is the data explosion that translates into cell towers, fiber cable, and data storage. To implement 5G networks, for example, a huge investment will be required in equipment, sites, and densification. The second is mobility, encompassing everything from vehicle design and energy powering cars and trucks, to controlled throughfares, shared usage, and sensing devices to manage and control urban infrastructure. Third is clean energy production and storage, driven by climate change and impending net-zero mandates. On the horizon are innovations affecting improvements in food production, water quality, increasing construction productivity, reducing building emissions, material science, health care, education, and urban infrastructure (Figure 23.2).

A green opportunity

In June 2022, a consortium of four large institutional investors came together to fund Haddington Venture's bid to construct the first of a series of green hydrogen platforms in western U.S. (O'Brien, 2022). This is a $650 million equity syndication funding the development of an underground salt cavern in Delta, Utah, set to become the site of the world's largest green hydrogen platform to date. Investors have the collective option to increase their investment

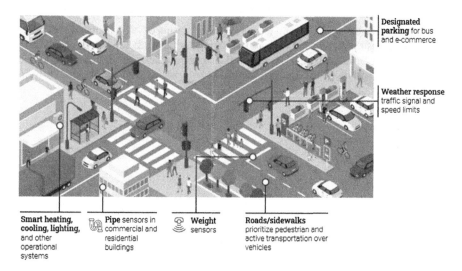

Figure 23.2 Digital potential in the built environment.

Source: Brookings Metro, Rebuild with purpose: An affirmative vision for 21st century American infrastructure.

to $1.5 billion. This development is a joint venture between Mitsubishi Power America and Magnum Development. The project also received a loan guarantee of $504.4 million from the U.S. Department of Energy. The Department of Energy has set aside $8 billion to support eight to 10 hydrogen hubs through the Biden infrastructure bill. The project will have the capacity to produce 100 metric tonnes per day, equivalent to 220 MW of electrolyze capacity, of green hydrogen under a long-term contract with the Intermountain Power Agency (IPA) in Utah. Haddington Ventures has historically invested in oil and gas storage facilities, and this is a first for them. Completion is scheduled for 2025.

This is an example of the rapid turnaround of a well-established firm in the oil and gas sector to capitalize on the ascendancy of green energy. The managing director of Addington commented, "When we started this initiative two and a half years ago, the traditional institutional investors largely misunderstood the opportunities that green hydrogen provided the renewable market." This project illustrates the combination of innovation and technical talent with investor and government support in a consortium arrangement in expectation of a significant swing to green hydrogen energy. It also reinforces the ability of sound technology to attract investor interest, mitigate some of the risks of a foray into new territory with a government-backed loan, and the crafting of a business model to incorporate a set of aligned business interests. Not to be lost is the fact that of the four large institutional investors, three are Canadian pension funds, and the fourth is GIC, Singapore's sovereign wealth fund. Globalization at its best.

Note

1 Levittown is the name of several large suburban housing developments created in the United States by William J. Levitt and his company Levitt & Sons. Built after World War II for returning white veterans and their new families, the communities offered attractive alternatives to cramped central city locations and apartments. https://en.wikipedia.org/wiki/Levittown

Conclusion
Now and into the future

This book focuses on a recent phenomenon that has global implications and touches upon the health of the planet, its people, and meeting future infrastructure needs. Private investment capital into infrastructure has historical precedents but a recent trend started with the new century and accelerated after the Great Financial Crisis of 2008. It is a trend marked by significant flows of private investment capital into infrastructure, a trend still in its naissance.

Academics and researchers have looked at *investment* spending on businesses to increase their competitiveness and productivity, and at government undertakings to support the needs of society and as a foundation for economic growth. In doing so, governments and industry emphasize the importance of investing in social and economic infrastructure to achieving their respective goals. Economists rely upon measures related to GDP growth to summarize levels of investment in infrastructure and, with a few exceptions, these measurements indicate a decline in the levels of public investment in infrastructure from a peak, post WWII. However, these measurements disguise a corollary trend; as governments reduce their investment levels, the capital markets are moving in the opposite direction and continue to commit substantial amounts of private capital to infrastructure. Little has been done to fully understand where these investment flows originate and why, what infrastructure they are directed to, plus understanding the various players in an industry with a complex network of capital flows, conduits, participants, and investment objectives.

When government officials state a need to increase the flow of private capital into infrastructure, this is a somewhat misguided pronouncement that reflects a woeful lack of understanding of what is happening in the private markets. Private capital is already flowing in large amounts into infrastructure, but not necessarily into the types of infrastructure that governments would like to see to address their specific needs or political ambitions. A significant disconnect between the public and the private sectors exists across the globe when it comes to understanding what drives private investments in infrastructure.

History tells us that the dichotomy between the public and private provision of infrastructure is not new. The pendulum has swung back and forth on

DOI: 10.1201/9781003396949-31

the ownership of infrastructure over decades and even centuries. It is not that one side, versus the other, is the more capable owner and better at serving the public interest. These swings reflect events of the times, the vagaries of political masters, the maturity of the capital markets, prevailing ideologies, and even military undertakings, conquests, and colonization efforts. Many private initiatives are based on monopolies that governments control and choose to leverage by selling exclusive rights to private parties. Politicians have always played a big role in deciding what the private sector can and cannot be involved in when it comes to infrastructure. There is no apparent logic behind many of their decisions and, as a result, investors have been cautious and even guarded whenever an opportunity to engage with the public sector arises. Aligning the interests of the public and the private sectors to deliver infrastructure is never an easy task.

By researching complex private investment undertakings in infrastructure, the book attempts to bridge two silos that appear to share little in common, other than believing that infrastructure is the foundation upon which society thrives and businesses flourish. The magnitude of this divide is represented by the "infrastructure gap" referred to in the book. This gap is evidence of the cost to society and business of the current predicament whereby governments have diminishing financial capacity to address their infrastructure needs in comparison to investors that face limited opportunities to invest their capital in infrastructure.

Local governments own much of the poorly maintained and outdated infrastructure and are best positioned to comprehend the infrastructure needs of their citizens. Most local governments lack the expertise and the financial capability to address a wide array of infrastructure needs from retrofitting existing infrastructure to commissioning new infrastructure. They also tend to be more cautious in engaging with the private sector in formal long-term business arrangements. This is despite the widespread involvement of private companies within urban areas in delivering infrastructure services that range from the provision of gas and electricity to garbage collection, Internet, communications, and data services.

Globally, the largest owner of infrastructure, when not restricted to considering only public infrastructure, is the private sector. That share is increasing as the public share diminishes. Compounding the problem is the divide between investing long term in existing infrastructure, the preference of private investors, and short-term investing in new infrastructure, the preference of public officials. When governments launch an appeal to increase private investment in infrastructure, they are implicitly referring to new infrastructure projects, often with a risk profile to which investors do not react well. Governments also dwell on infrastructure procurements models rather than operational models, with the exception of public–private partnerships, with which most local governments have very limited experience.

Governments are prone to refer to private investors as if they are a homogeneous lot, which they are not. They also don't distinguish between the financing and the funding of infrastructure and often infer that investors are a potential source of financing (debt), much like a commercial bank. Governments will focus on matters of public policy, cost-efficient distribution of services, social equity, public finance, and future economic growth. Investors focus on the ability to apply equity to generate targeted returns, risk mitigation, and build a sustainable enterprise.

Governments refer almost exclusively to assets that are physical or tangible— things that can be seen and touched. Tangible infrastructure is the basis of measuring the infrastructure gap. On the other hand, investors are increasingly attracted to the opportunities to invest in non-tangible infrastructure assets. Until very recently, the investments that national statistical offices measured were all tangible assets. This is no longer the case as surveys and analysis now depict the size and scope of intangible investments that have been growing in almost all developed countries and in some countries now outweighs tangible investments. Intangible infrastructure assets add a totally new dimension to infrastructure investing that the public realm is just beginning to acknowledge.

Hopefully, this book clarifies some of the issues raised. The intent is to encourage a search for solutions based on a better understanding of the challenges involved. There is an urgent need for the two sides to find ways to turn the corner on closing the infrastructure gap. It must be recognized that the gap is likely to widen given the consequences of COVID, the war in Ukraine, predictions of an impending recession, and the growing uncertainty about the global geo-political situation. For many governments the cupboard is bare. If needs are not met, the scale and magnitude of the infrastructure problem will exacerbate the political instability and the social inequity that are growing across large parts of the globe. Infrastructure is about more than building new energy plants, toll roads, and cell towers. It is about clean water, access to food, health care, security, and education for all. Private capital has a role to play in meeting many of these needs.

Key arguments in the book are summarized in the following points:

1   There is no universally agreed-upon definition of what constitutes infra-structure. Infrastructure, as a term, generally refers to tangible things that can be seen and touched. This limitation no longer applies. Economists and governments rely on definitions that embrace a public interest and can be incorporated into public accounts. Investors use the term to broadly describe an asset class in which they choose to invest, and the reference can embrace a wide array of assets. Investors have limited interest in defining infrastructure. Infrastructure is whatever meets their investment criteria and the term captures an ever-expanding set of assets. The differences in describing what constitutes infrastructure has a significant impact on data collection and analysis, as well as policy frameworks, and reinforces the divide between the public and the private sectors.

2   The much-quoted "infrastructure gap" is a theoretical construct framed in a manner to remind government officials and their political masters

of the cumulative implications of their inability to address infrastructure needs. It has little consequence for investors, other than to inform them of the magnitude of funding that governments will need if they ever choose to address this gap. There is scant evidence, and despite the usual rhetoric about involving private funds, that closing the gap is going to happen anytime soon. Governments don't have the funds, will be even poorer in the post-COVID era, and have not come forward with investment models, beyond PPPs, to engage the private sector. But the infrastructure gap is about more than just money. Impediments to closing the gap, in addition to lack of funds, include issues on ownership, political ideology, fear in many countries about privatizing what many consider to be a "free" good or service, and the lack of business models that can engage both sectors and "squeeze in" private capital.

3   While public investment in infrastructure has been declining for many years, with a few exceptions, the opposite is true of private investment in infrastructure. What one investor refers to as a *"wall of money"* is a result of significant year-to-year increases in investment capital committed to acquiring infrastructure assets that started back in 2000 and accelerated post 2008. The global competition to acquire infrastructure assets is intense, driving prices upward and driving down returns. Each year the amount of *"dry powder"* increases, an amount that represents the total of private funds committed, but not yet allocated. For investors, the common refrain is too much money and too few investment opportunities. The challenge that private investors face is not raising money but spending money.

4   Despite what many think, including government officials, the largest portion of infrastructure that society depends upon is privately owned. A significant portion of this ownership resides within private corporations and is difficult to account for. Private ownership of infrastructure is not new and dates back centuries. In many situations, where growth was an imperative, it was politically expedient for governments to rely exclusively on the private sector to provide much of their needed infrastructure from railroads, canals, and seaports to energy and communications. The big swing to public ownership of infrastructure occurred post-WWII and the swing back again to the private sector began in the Thatcher era. The debate on public versus private infrastructure is as heated today as it ever was. This debate reflects ideological positioning, the vagaries of political masters, historical precedents, the financial health of governments, and even responses to emerging technologies. Privatization of infrastructure is seen in many developed societies, including the U.S., as a direct threat to the public interest. Critics will claim that privatization represents a subjugation of the public interest to entities whose sole purpose is to make money. Unfortunately, it is debate that hinders progress in bridging the two divides.

5   Public officials see infrastructure primarily through two lenses: the need to address deferred maintenance or outdated infrastructure; and the need for new infrastructure to meet new or growing needs. Governments focus

on procurement, short time frames, and projects that can generate political goodwill. Investors see infrastructure as a business opportunity that can reap rewards over long time periods, and for which they can manage capital requirements. Public officials focus on the legitimacy of procurement processes that involve large capital expenditures with the potential to deliver on budget and on time. Investors focus on the functionality and efficiency of operations and on the human resources that support these operations. As one major pension fund investor succinctly stated, *"in acquiring infrastructure, our focus is on people first, projects second"*.

6   Investment decisions are made in a portfolio context comprised of three "buckets"—real assets, fixed income, and equities. The relative weighting of each provides an overall net return sufficient to meet funding obligations. Infrastructure falls into the "real asset" bucket and has unique characteristics that enhance its diversification attributes and its appeal to long-term investors. The concept of real or tangible assets contributing to an inflation-sensitive portfolio has been gaining ground in the last decade and a portfolio mix of real estate, infrastructure, and natural resources is particularly popular among large institutional investors. The governing factors in all investment decisions are the long-term obligations to fund pensions, insurance claims, and match the return expectations of fund investors. In recent years, investments in infrastructure have played a significant role in meeting these obligations. In a portfolio context, and based on recent performance, allocations to infrastructure are likely to continue to increase relative to the other asset classes.

7   Investors have a clear preference for mature markets—secondary transactions and brownfield projects, with investments in energy, transportation, and communications. Where possible, they prefer domestic markets but are expanding their global presence. They have largely avoided the emerging markets and frontier economies but realize this is where future opportunities reside. They also avoid greenfield projects including public–private partnerships, but for different reasons. Greenfield projects introduce risks, such as construction risk, that investors are not equipped to deal with. Nor is there the immediate prospect of a return on what can be a sizeable capital outlay for 5 or even 10 years until a project is delivered and operational. PPPs have a limit capacity for equity contributions, are legally very complex, and many are based on availability-based payments that offer bond-like returns. Many investors claim that PPPs are structured to intentionally "squeeze out" rather than squeeze in private capital.

8   The flow of investment capital from where sourced to where allocated represents a complex web of investment structures, firms, and risk-return profiles. This network includes listed funds, unlisted funds, and direct ownership. Tracking the flow of funds across this complex web is not an easy task, particularly given that the unlisted funds represent the largest amount of investment capital and lack transparency as they operate as fee-driven private entities, with limited obligations to report on their activities.

The unlisted funds use return data in their marketing programs to calculate performance-based fees, guide their investment allocations, and calculate incentive-based payments to employees. Institutional investors, and particularly pension funds and insurance firms, report annually to their beneficiaries and pensioners. Listed funds are priced daily on exchanges and offer liquidity, at a price. The problem is that there are no reliable benchmarks for infrastructure investing as empirical data are hard to access. In the absence of reliable data, the question remains: Is infrastructure delivering the cash flows and returns to match the claims that are publicized?

9   If governments and the private sector choose to work together to increase the flow of private capital into new infrastructure projects and make a dent in the infrastructure gap, new business models will be required. Business models have always been at the core of any appeal to engage the private sector as evidenced by the many historical precedents that shaped the infrastructure we rely upon today. From the 1990s onward, the term "business model" in the private sector has been used to describe innovative ways of practicing or doing business with an emphasis on data handling, communication technologies, the Internet, and e-commerce. There is no one definition of a business model, or a template to follow. However, there is an understanding that business models are built around a value proposition with the potential to generate and sustain revenues. For monies to flow from investors into infrastructure projects jointly undertaken with governments, two things are required: first, governments must adopt a way of thinking, understand common business terminology, and appreciate the core logic driving a value proposition; and second, governments must be willing to identify sources of revenue to drive the value proposition.

10   Investors realize that as much as 75 percent of the infrastructure required globally over the next 30 years has yet to be built, or even imagined in many sectors. Investors cannot ignore the potential of the market for new projects. Emerging technologies are already reshaping the infrastructure investment market and attracting investors to greenfield projects in an array of business arrangements, most of which have limited direct public sector involvement. Possibilities range from breakthroughs in artificial intelligence, electric vehicles, robotics, energy production and storage, and data-driven platforms. The downsides are the risks in introducing anything new, the resulting obsolescence of existing infrastructure, and the ability of governments to update policies and regulations, recognizing that infrastructure investments often require close collaboration and support from governments. Just as infrastructure itself will be hindered by the human factor—lack of a skilled workforce—this will be even more severe among government departments and agencies involved in infrastructure, despite the best of intentions.

11   In addition to emerging technologies, the investment landscape is being reshaped by investments in non-tangible infrastructure assets.

Opportunities to invest in intangible infrastructure will require flexibility, willingness to take new risks in a "trial and error" implementation framework, new forms of measurement and assessment, and new management tools. It may also result in new ways of working, in different locations, and with associated changes in human behavior that are difficult to predict. Intangibles will create new markets, larger scale operations, and give rise to new synergies, new partnerships and collaborations, and different rewards systems. These are not insurmountable challenges, nor are they going to slow the trend to increased investment in intangibles.

12 Whether adapting existing infrastructure or building new infrastructure, investors must embrace sustainability in its many dimensions. This is not about "greening" our infrastructure, or meeting ESG requirements. Sustainability in the broader context refers to contributing to climate change initiatives, enhancing the resilience of the built environment, addressing social inequities, addressing price distortions that destabilize markets, or doing more with less. Sustainability not only makes good business sense, but should be an integral part of enterprise risk management to protect and enhance the value of investments over long time periods. While the benefits of sustainable infrastructure are long term, there is an immediate urgency to take advantage of a very small window of opportunity, perhaps as short as the next decade, through which to launch critical sustainable infrastructure initiatives.

The need for infrastructure is driven by growth. What is predictable is that population growth, the growth of cities, growth in the emerging economies, and the aging of the population in the developed world will continue. This pace of growth will not be matched in developed countries with the exception of aging, Japan being the often-cited example. What are not easy to predict are the consequences of the geo-political uncertainty that is increasingly prevalent across global markets. Infrastructure does best when countries cooperate, political and economic links are strengthened, economies prioritize the needs of people, and economic and social prosperity thrives.

What we have today is the return of great-power competition, imperial ambitions, fights over resources, and the large-scale displacement of people (Haass, 2022). Add to this list the pandemic, climate change, nuclear proliferation, war, and the prospect of a global recession. Countries are no longer working together on regional and international challenges. Problems are compounded by the rise of social disorder, political instability, and nationalism. Investments in infrastructure are part of the machinery of dealing with societal challenges and the global community will suffer if investments in infrastructure dry up in the decades ahead.

The use of private capital to fund infrastructure has deep historical roots. In the post-WWII era, industry and governments in the mature economies had the luxury of a broad range of policy choices and access to alternative sources of public and private funds. Despite this freedom to choose, many governments chose to do as little as possible in addressing obvious infrastructure needs. Some chose to do nothing. Two obvious consequences are the infrastructure gap and climate change at an accelerated pace. What might be a revelation to some is the realization that many of the infrastructure challenges will shift to the emerging and frontier economies. It will be a handful of countries, whether in developed or emerging economies, that will find a way to engage with the private sector and harness the massive amounts of private capital to address their infrastructure needs to fuel their prosperity. Hopefully, this book has brought about a better understanding of the role of private capital and the emergence of a flourishing global industry that can play a major role in funding infrastructure—a role that is still not widely understood.

References

Airoldi, Marco, Jeffrey Chua, Philip Gerbert, Jan Justus, and Raffael Rilo (2013). *Bridging the Gap: Meeting the Infrastructure Challenge with Public-Private Partnerships.* February 2013. www.utap.gov.pt/Publicacoes_interna/Bridging_the_Gap_Feb_2013_tcm80-128534.pdf

Akintoye, Akintola, Matthias Beck, and Mohan Kumaraswamy (Eds). (2015). *Public Private Partnerships: A Global Review.* Routledge.

Alonso, Javier, Alfonso Arellano, and David Tuesta (2015). *Pension Fund Investment in Infrastructure and Global Financial Regulation.* Pension Research Council Working Paper Pension Research Council. The Wharton School, University of Pennsylvania. https://repository.upenn.edu/prc_papers/82/

Alves, Bruno. (2020). The Decade. *Infrastructure Investor.* Special Edition. July 2020. www.infrastructureinvestor.com/download-the-infrastructure-investor-10th-anniversary-issue/

Amanya, Franklin S. and Gitahi Njenga, G. (2022). Construction delays and project cost overrun: a case of regional cybercrime center in Gasabo District, Rwanda. *Journal of Entrepreneurship & Project Management.* 6 (1): 34–52. https://stratfordjournals.org/journals/index.php/journal-of-entrepreneurship-proj/article/view/1072

Andonov, Alexander, Roman Krausl, and Joshua D. Rauh (2021). *Institutional Investors and Infrastructure Investing.* Stanford University Graduate School of Business. Research Paper No. 18-42.

Athukorala, Rohantha Neville Anthony, Do Ba Khanga, Anurug Ruangrob, Malek Ajdar Barzegar Sedigh, THV Mothilal De Silva, Vinod Guota and I.M. Pandey (2017). Managing large projects in emerging markets. *Vilapa* 32 (3): 16. www.deepdyve.com/lp/sage/managing-large-projects-in-emerging-markets-JOfuMQbdUC

AUDITBOARD. (2018). What is operational risk management? *The Overview.* www.auditboard.com/blog/operational-risk-management/

Australian Government, Department of Infrastructure and Regional Development. (2015). *National Alliance Contracting Guidelines.* National Alliance Contracting Guidelines–Guide to Alliance Contracting (infrastructure.gov.au). www.infrastructure.gov.au/sites/default/files/migrated/infrastructure/ngpd/files/National_Guide_to_Alliance_Contracting.pdf

Aviva Investors. (2020). *Intangible Infrastructure: Storing and Transmitting Data.* www.avivainvestors.com/en-gb/views/aiq-investment-thinking/2020/02/intangible-infrastructure-storing-and-transmitting-data/

Baba, Razak Musah. (2022). Ontario teachers, Corio team up to develop offshore wind projects. *IPE Real Assets*. May 12, 2022. Ontario Teachers, Corio team up to develop offshore wind projects | News | Real Assets (ipe.com)

Barlow, James and Martina Köberle-Glaiser. (2009). Delivery innovation in hospital construction: contracts and collaboration in UK's private finance initiative hospital program. *California Management Review* 51(2): 126–143.

Beckers, Frank and Uwe Stegeman. (2021). *A Smarter Way to Think about Public-Private Partnerships*. McKinsey & Company. www.mckinsey.com/capabilities/risk-and-resilie nce/our-insights/a-smarter-way-to-think-about-public-private-partnerships

Beckers, Frank, Nicola Chiara, Adam Flesch, Jiri Maly, Eber Silva, and Uwe Stegman. (2013). *A Risk-Management Approach to a Successful Infrastructure Project*. McKinsey Working Papers on Risk, No.52.

Bédard-Pagé, Guillaume, Annick Demers, Eric Tuer, and Miville Tremblay. (2016). Large Canadian public pension funds: A financial systems perspective. *Financial Systems Review*. Bank of Canada. www.bankofcanada.ca/wp-content/uploads/2016/ 06/fsr-june2016-bedard-page.pdf

Bennett, Jennifer, Robert Kornfeld, Daniel Sichel, and David Wasshausen. (2020). Measuring infrastructure in BEA's National Economic Accounts. *National Bureau of Economic Research* (NBER). Draft paper.

Bentley, Zac. (2022a). Court strikes down Macquarie-led PPP in Pennsylvania. *Infrastructure Investor*. July 7, 2022.

Bentley, Zac. (2022b). How the world's largest infra-allocator is supersizing its portfolio. *Infrastructure Investor*. July 7, 2022.

Betz, Eric. Aqueducts. (2020). How Ancient Rome brought water to its people. *Discover Magazine*. October 26, 2020. www.discovermagazine.com/planet-earth/aqueducts-how-ancient-rome-brought-water-to-its-people

Bhattacharya, Amar, Mattia Romani, and Nicholas Stern. (2012). *Infrastructure for Development: Meeting the Challenge. Policy Paper*. London: Grantham Research Institute on Climate Change and the Environment and Centre for Climate Change Economics and Policy, London School of Economics and Political Science. www.lse. ac.uk/GranthamInstitute/wp-content/uploads/2014/02/PP-infrastructure-for-deve lopment-meeting-the-challenge.pdf

Bhattacharya, Amer, Joshua P. Meltzer, Jeremy Oppenheim, Zia Qureshi, and Nicholas Stern. (2016). *Delivering on Sustainable Infrastructure for Better Development and Better Climate*. Brookings: Global Economy and Development. www.brookings.edu/ research/delivering-on-sustainable-infrastructure-for-better-development-and-bet ter-climate/

Bielenberg, Aaron, Mike Kerlin, Jeremy Oppenheim, and Melissa Roberts. (2016). *Financing Change: How to Mobilize Private Sector Financing for Sustainable Infrastructure*. McKinsey & Company. https://newclimateeconomy.report/workin gpapers/wp-content/uploads/sites/5/2016/04/Financing_change_How_to_mobil ize_private-sector_financing_for_sustainable-_infrastructure.pdf

Blecher, Lennard, Alex Darden, and Matthias Fackler. (2019). The decade. *Infrastructure Investor*. Special Edition.

Bloom, Jerry R. (2015). *An Historic View of Clean Energy Power Purchase Agreements*. Winston & Strawn LLP. Power Point presentation. Washington, DC. https://provost. gwu.edu/historical-overview-ppa

Bloomberg. (2021). Muckesh Ambani's $50 phone built by Google can unleash a credit revolution for banking. *Times of India*. September 3, 2021. https://timesofindia.ind iatimes.com/business/india-business/mukesh-ambanis-50-phone-built-by-google-can-unleash-a-credit-revolution-for-banks/articleshow/85896318.cmsBouwer, Jap, Vik Krishnan, Steve Saxon, and Carline Tufft. (2022). *Taking Stock of the Pandemic's Impact on Airports*. McKinsey & Company. March 2022. www.mckinsey.com/industr ies/travel-logistics-and-infrastructure/our-insights/taking-stock-of-the-pandemics-impact-on-global-aviation

Brilha, Nuno Mocica and Helena Nobre. (2019). Airports as platforms: towards a new business model. *International Journal of Business Performance Management* 20(4):297.

Brinkman, Marcel and Vijay Sarma. (2022). *Infrastructure Investing Will Never be the Same*. McKinsey & Company. www.mckinsey.com/industries/private-equity-and-principal-investors/our-insights/infrastructure-investing-will-never-be-the-same

Brookfield Asset Management. (2016). *Real Assets, Real Diversification.*

Brookings Institute and the Urban Institute. (2022). *What is the Highway Trust Fund, and how is it financed?* The Tax Policy Center's Briefing Book. www.taxpolicycenter. org/briefing-book/what-highway-trust-fund-and-how-it-financed

Brown, Aran. (2022). Investor funds and investor report 2021. *IJInvestor*. www.ijglobal. com/articles/161570/ijinvestor-funds-and-investors-report-2021

Bryson, John, Andy Pike, Claire Walsh, Tim Foxon, and Chris Bouch. (2014). Infrastructure Business Models (IBM) Working Paper. *Ibuild*, Briefing Note No. 2. https://research.ncl.ac.uk/media/sites/researchwebsites/ibuild/BP2%20-%20Infrast ructure%20business%20model%20definition_DRAFT.pdf

Buck, Martin. (2017). *Crossrail Project: Finance, Funding and Value Capture for London's Elizabeth Line*. Institute of Civil Engineers. ICE Proceedings. www.icevir tuallibrary.com/doi/10.1680/jcien.17.00005

Bult-Spiering, Mirjam, Geert Dewulf, and Annoleous Balnken. (2008). *Strategic Issues in Public-Private Partnerships: An International Perspective*. Wiley.

Campbell, Drew. (2022). The money talks. *Institutional Investing in Infrastructure* 15(1). https://irei.com/publications/article/the-money-talks-the-i3-editorial-advisory-board-met-in-california-to-discuss-infrastructure-investing/

Cassidy, John. (2022). The averted national rail strike is a parable of contemporary American capitalism. *The New Yorker*. www.newyorker.com/news/our-columni sts/the-averted-national-rail-strike-is-a-parable-of-contemporary-american-cap italism

Center for Climate and Energy Solutions. (2022). *Climate-Related Financial Risk*. 2022. www.c2es.org/content/climate-related-financial-disclosures/

Charbonneau, M. (2012). New public management. In: L. Côté and J.-F. Savard (Eds.) *Encyclopedic Dictionary of Public Administration*. www.dictionnaire.enap.ca

Chase, Melisa. (2021). *Simplifying Sustainable Finance-Explaining Green Bonds, Green Loans, Sustainability-linked loans and Bonds and More*. Morningstar Analytics. www. sustainalytics.com/esg-research/resource/corporate-esg-blog/simplifying-sustainable-finance-green-loans-vs-green-bonds-vs-sustainability-linked-loan-and-more

Chattopadhyay, Soumyadip. (2006). Municipal bond market for financing infrastruc- ture. *Economic and Policy Weekly* 41(26): 2787–2791. www.epw.in/journal/2006/26/ special-articles/municipal-bond-market-financing-urban-infrastructure.html

ChemEurope.Com. (2022). *Sanitation in Ancient Rome*. www.chemeurope.com/en/encyclopedia/Sanitation_in_ancient_Rome.html

Cheong, Teo Eng. (2015). *World Bank Blogs*. November 2015. https://blogs.worldbank.org/team/teo-eng-cheong

Christensen, Tom and Per Laegreid. (2011). *Post-NPM Reforms: Whole Government Approaches as a New Trend. New Steering Concepts in Public Management.* Chapter 2. October 2011. www.emerald.com/insight/publication/doi/10.1108/S0732-1317(2011)21

Christensen, Clayton, Thomas Bartman, and Derek van Bever. (2016). The hard truth about business model innovation. *MIT Sloan Management Review* 58(1), Fall 2016. https://sloanreview.mit.edu/article/the-hard-truth-about-business-model-innovation/

Cisneros, Henry. (2010). America's essential infrastructure: A key to competitiveness. *The Handbook of Infrastructure Investing,* Michael Underhill (Ed.). Wiley. www.wiley.com/en-us/The+Handbook+of+Infrastructure+Investing-p-9781118268117

Climate Policy Initiative (CPI). (2021). *Global Landscape of Climate Finance 2021*. www.climatepolicyinitiative.org/publication/global-landscape-of-climate-finance-2021/

Coalition for Climate Resilient Investment (CCRI). (2022). *Guidelines for Integrating Physical Climate Risks in Infrastructure Investment Appraisal: The Physical Climate Risks Assessment Methodology (PCRAM)*. Embargoed Version. 2022.

Convergence. (2022a). *Blended Finance*. www.convergence.finance/blended-finance#deal-sizes-and-types

Convergence. (2022b). *State of Blended Finance*, 2022, Climate Edition. www.convergence.finance/resource/state-of-blended-finance-2022/view

Convergence. (2023). *Blended Finance*. www.convergence.finance/blended-finance#definition

CPP Investments. (2021). *Annual Report 2021*. www.cppinvestments.com/the-fund/our-performance/financial-results/f2021-annual-results

Croce, Raffaele Della and Stefano Gatti. (2014). *Financing infrastructure –International trends. OECD Journal of Financial Market Trends.* www.oecd.org/daf/fin/financial-markets/Financing-infrastructure-international-trends2014.pdf

Cutler, David and Grant Miller. (2005). The role of public health improvements in health advances: the twentieth-century United States. *Demography* 42(1):1–22. https://pubmed.ncbi.nlm.nih.gov/15782893/

Czigler, Thomas, Sebastian Reiter, Patrick Schultz and Ken Somers. (2020). *Laying the Foundations for Zero-Carbon Cement*. McKinsey & Company. May 2020. www.mckinsey.com/~/media/McKinsey/Industries/Chemicals/Our%20Insights/Laying%20the%20foundation%20for%20zero%20carbon%20cement/Laying-the-foundation-for-zero-carbon-cement-v3.pdf

Darwin, John. (2013). *Unfinished Empire; The Global Expansion of Britain*. Penguin Books.

Davenport, Romola Jane, Max Satchell, and Leigh Matthew William Shaw-Taylor. (2019). *Cholera as a 'Sanitation Test' of British Cities,* 1831–1866. National Library of Medicine. www.ncbi.nlm.nih.gov/pmc/articles/PMC6582458/

David, Paul A. (1990). The dynamo and the computer: An historical perspective on the modern productivity paradox. *American Economic Review* 80(2):3551–3561. www.jstor.org/stable/2006600

Davies, Andrew, Lars Fredericksen, and Geert Dewulf. (2010). *Business Models, Infrastructure and the Changing Public-Private Interface.* EPOS Working Paper Proceedings.

Deep, Akash. (2022). Infrastructure finance. Jose A. Gomez-Ibanez and Zhi Liu (Eds.) *Infrastructure Economics and Policy: International Perspectives*. Lincoln Institute of Lan Policy. Chapter 9, 213–238.

Delmon, Jeffrey. (2011). *Public-Private Partnership Projects in Infrastructure 2nd Edition*. Cambridge University Press.

Deutsche Asset Management. (2017). *Why Invest in Infrastructure?* Research Report. May 2017. www.dws.com/globalassets/institutional/research/pdfs/Deutsche_AM_Why_Invest_in_Infrastructure_May_2017.pdf

Dobbs, Richard, James Manyika, Jonathan Woetzel, Jaana Remes, Jesko Perrey, Greg Kelly, Kanaka Pattabiraman, and Hemant Sharma. (2016). *Urban World: The Global Consumers to Watch*. McKinsey Global Institute. www.mckinsey.com/featured-insights/urbanization/urban-world-the-global-consumers-to-watch

Drucker, Peter F. (1954). *The Practice of Management*. Harper & Row. New York.

Dunn, Rob, Adi Kumar, Nehal Mehta, Sara O'Rourke, and Tim Ward. (2022). *A New Era of US Infrastructure Grants*. McKinsey & Company. www.mckinsey.com/industries/public-and-social-sector/our-insights/a-new-era-of-us-infrastructure-grants

Economist. (2021). Was COP26 in Glasgow a success? November 14, 2021. www.economist.com/international/2021/11/14/was-cop26-in-glasgow-a-success

Economist. (2022a). *ESG should be boiled down to one simple measure: emissions*. July 21, 2022. www.economist.com/leaders/2022/07/21/esg-should-be-boiled-down-to-one-simple-measure-emissions

Economist. (2022b). *Keeping trains apart is crucial to safety*. September 28, 2022. www.economist.com/science-and-technology/2022/09/28/keeping-trains-apart-is-crucial-to-safety .

Economist. (2022c). The construction industry remains horribly climate unfriendly. June 15, 2022. www.economist.com/finance-and-economics/2022/06/15/the-construction-industry-remains-horribly-climate-unfriendly

Economist. (2022d). The property industry has a huge carbon footprint. Here's how to reduce it. June 15, 2022. www.economist.com/leaders/2022/06/16/the-property-industry-has-a-huge-carbon-footprint-heres-how-to-reduce-it

Economist. (2022e). Today's heatwaves are a warning of worse to come. July 20, 2022. www.economist.com/leaders/2022/07/20/todays-heatwaves-are-a-warning-of-worse-to-come

EDHEC Infrastructure Institute. (2021). *Towards a Scientific Approach to ESG for Infrastructure Investors*. March 2021. https://edhec.infrastructure.institute/wp-content/uploads/2021/03/ESG_Approach_Roadmap_2021.pdf A

Edwards, Chris. (2017). *Who Owns U.S. Infrastructure?* Tax and Budget Bulletin No. 78. CATO Institute. www.cato.org/tax-budget-bulletin/who-owns-us-infrastructure

Ehlers, Torsten. (2014). *Understanding the Challenges of Infrastructure Finance*. Bank for International Settlement. Working Paper 454. www.bis.org/publ/work454.htm

Elliot, Larry. (2019). World economy is sleepwalking into a financial crisis, warns Mervyn King. *The Guardian*, October 20, 2019. www.theguardian.com/business/2019/oct/20/world-sleepwalking-to-another-financial-crisis-says-mervyn-king#:~:text=%E2%80%9CAnother%20economic%20and%20financial%20crisis,are%20sleepwalking%20towards%20that%20crisis.%E2%80%9D

Emery, Herb. (2015). *A Brief History of Infrastructure in Canada, 1870–2015*. Department of Economics, School of Public Policy, University of Calgary. Slide presentation. www.queensu.ca/iigr/sites/iirwww/files/uploaded_files/EmeryHerbSOTF2015.pdf

Engel, Eduardo, Ronald D. Fischer, and Alexander Galetovic. (2020). *When and How to Use Public-Private Partnerships in Infrastructure: Lessons from International Experience.* National Bureau of Economic Research. Working Paper 26766. www. nber.org/papers/w26766

European Investment Bank. (2010). *Public and Private Financing of Infrastructure. EIB Papers* 15(1). www.eib.org/attachments/efs/eibpapers/eibpapers_2010_v15_n02 _en.pdf

Fabre, Anaïs, and Stephen Straub. (2021). *The Impact of Public-Private-Partnerships (PPPs) in Infrastructure, Heath, and Education.* Toulouse School of Economics. Working Paper No. 986. September 2021. https://publications.ut-capitole.fr/id/epr int/30926/1/wp_tse_986.pdf

Fenn, Michael. (2022). *More Affordable Infrastructure: Tax-Free Municipal Bonds.* Strategy Corp. https://strategycorp.com/2022/08/more-affordable-infrastructure-tax-free-municipal-bonds/

Fidelity Investments. (2022). *SRI vs. ESG vs. Impact Investing: What's the Difference?* www.fidelity.ca/en/investor/investorinsights/srivsesgvsimpactinvesting/#:~:text= Environmental%2C%20social%20and%20corporate%20governance,to%20soci ety%20(or%20both).

Finnerty, John D. (2013). *Project Financing: Asset-Based Financial Engineering.* Wiley.

Foster, Kendrick. (2021). *Building (and Canceling) an Airport for Mexico City.* Harvard Political Review. April 12, 2021. https://harvardpolitics.com/mexico-city-airports/

Frampton, Stepehen. (2023). *Brookfield: Transparently Bullish Analysis so You Can Decide for Yourself.* Seeking Alpha. January 18, 2023. https://seekingalpha.com/aut hor/stephen-frampton

Frigo, Mark L. and Richard J. Anderson. (2011). What is strategic risk management? *Strategic Finance.* April 2011. www.markfrigo.org/What_is_Strategic_Risk_Mana gement_-_Strategic_Finance_-_April_2011.pdf

Gallas, Danial. (2019). *Brazil's Odebrecht Corruption Scandal Explained.* BBC South America. April 17, 2019. www.bbc.com/news/business-39194395

Garcia, Moraleja Silvia and Tim Whittaker. (2019). *ESG Reporting and Financial Performance: The Case of Infrastructure.* EDHECinfra. March 2019. https://edhec. infrastructure.institute/wp-content/uploads/2019/03/Garcia_Whittaker_2019. pdfGaremo, Nicklas, Stefan Matzinger, and Robert Palter. (2015). *Megaprojects: The Good, the Bad, and the Better.* McKinsey & Company. July 2015. www.mckinsey.com/ capabilities/operations/our-insights/megaprojects-the-good-the-bad-and-the-better

Gatti, Stefano. (2008). *Project Finance in Theory and Practice.* Elsevier, Academic Press.

Gatti, Stefano. (2018). *Project Finance in Theory and Practice: Designing, Structuring, and Financing Private and Public Projects.* Academic Press.

George Inderst. (2021). *Financing Development: Private Capital Mobilization and Institutional Investors.* Inderst Advisory. https://papers.ssrn.com/sol3/papers. cfm?abstract_id=3806742

Gil, Nuno and Sara L. Beckman. (2009). Infrastructure meets business: Building new bridges, mending old ones—An introduction to the special edition. *California Management Review.* 51(2):17–18. https://journals.sagepub.com/doi/abs/10.2307/ 41166478?journalCode=cmra

Glaeser, Edward L. (2022). Infrastructure and urban form. In: Gómez-Ibáñez, José A. and Zhi Liu. (Eds.) *Infrastructure Economics and Policy: International Perspectives.* Chapter 8, 91–199.Lincoln Institute.

Glaeser, Edward L. and James M. Poterba (Ed). (2021) *Economic Analysis and Infrastructure Investment*. National Bureau of Economic Research, University of Chicago Press.

Global Infrastructure Hub and Oxford Economics. (2017). *Global Infrastructure Outlook*. https://cdn.gihub.org/outlook/live/methodology/Global+Infrastructure+Outlook+-+July+2017.pdf

Global Infrastructure Hub. (2021). *Infrastructure Monitor 2021*. www.gihub.org/resources/publications/infrastructure-monitor-2021-report/

Global Infrastructure Hub. (2022). *Data Insights*. www.gihub.org/infrastructure-monitor/insights/for-investors-seeking-to-diversify-and-optimise-their-portfolios-infrastructure-debt-and-unlisted-infrastructure-equities-are-very-strong-options-according-to-long-term-data/

Global Infrastructure Hub. 2019. *Leading Practices in Governmental Processes Facilitating Infrastructure Project Preparation*. www.gihub.org/resources/publications/leading-practices-in-governmental-processes-facilitating-infrastructure-project-preparation/

GlobalData. (2017). *Global Infrastructure Outlook*. www.globaldata.com/store/report/global-infrastructure-outlook/

Gómez-Ibáñez, José A. and Zhi Liu (Ed). (2022). *Infrastructure Economics and Policy: International Perspectives*. Lincoln Institute.

Goolsbee, Austan and Chad Syverson. (2022). *The Strange and Awful Path of Productivity in the U.S. Construction Sector*. National Bureau of Economic Research. www.nber.org/books-and-chapters/technology-productivity-and-economic-growth/strange-and-awful-path-productivity-us-construction-sector

Gordon, Robert J. (2016). *Rise and Fall of American Growth*. Princeton University Press.

Gramlich, Edward M. (1994). Infrastructure investment: A review essay. *Journal of Economic Literature* 32(3): 1176–1196.

Green, Nicole and Madeleine Chua. (2018). Alliance contracting: Key ingredients to a successful alliance. *Roads and Infrastructure*. April 15, 2018. https://roadsonline.com.au/alliance-contracting-key-ingredients-to-a-successful-alliance/

Greenstein, Shane. (2021). Digital infrastructure. In: *Economics of Infrastructure Investment*. University of Chicago Press. https://doi.org/10.7208/chicago/9780226800615-011

Gregg, Neil S. (2009). *Infrastructure Finance: The Business of Infrastructure for a Sustainable Future*. Wiley.

Grimsey, Darwin and Melvin Lewis. (2007). *Public Private Partnerships: The Worldwide Revolution in Infrastructure Provision and Project Finance*. Edward Elgar. Cheltenham, UK.

Haass, Richard. (2022). The dangerous decade: A foreign policy for a world in crisis. *Foreign Affairs*. Sep/Oct 2022. www.foreignaffairs.com/united-states/dangerous-decade-foreign-policy-world-crisis-richard-haass

Hahn, Devin and Amy Laskowski. (2002). *Tracing the History of New England's Ice Trade*. The Brink, Boston University. February 4, 2022. www.bu.edu/articles/2022/tracing-the-history-of-new-england-ice-trade/

Harchaoui, Tarek M., Faouzi Tarkhani, and Paul Warren. Public infrastructure in Canada: Where do we stand? *Statistics Canada*. Catalogue No. 11-624-MIE-No. 005. 2003. https://webdocs.edmonton.ca/InfraPlan/Infra/Reports/statscan_public-Infrastructure%20_canada.pdf

Haskel, Jonathan and Stain Westlake. (2018). *Capitalism Without Capital: The Rise of the Intangible Economy*. Princeton University Press. Princeton, NJ. P4

Hax, Arnoldo C. and Nicolas S. Majluf. (1996). *Strategy Concept and Process: A Pragmatic Approach*, (2nd Edition). Prentice Hall, Englewood, NJ.

Hazan, Eric, Sven Smit, Jonathan Woetzel, Biljana Cvetanovski, Mekala Krishnan, Brian Gregg, Jesko Perrey, and Klemens Hjartar. (2021). *Getting Tangible about Intangibles: The Future of Growth and Productivity*. McKinsey & Company. www.mckinsey.com/capabilities/growth-marketing-and-sales/our-insights/getting-tangible-about-intangibles-the-future-of-growth-and-productivity

Henry, Emily. (2019). Why the biggest isn't always the best. *The Decade*. Infrastructure Investor. www.infrastructureinvestor.com/the-decade-present/

Hodge, Graeme J., Carson Greve, and Anthony Boardman (Ed). (2015). *International Handbook on Public-Private Partnerships*. Edward Elgar Publishing.

Hodge, Graeme, Carston Greve, and Anthony Boardman. (2010). Introduction: the PPP phenomenon and its evaluation. *International Handbook of Public-Private Partnerships*, p. 3. Edward Elgar Publishing Limited. www.lse.ac.uk/granthaminstitute/wp-content/uploads/2014/03/PP-infrastructure-for-development-meeting-the-challenge.pdf

Hussain, Sherena and James McKellar. (2020). Exploring the success of social infrastructure public-private partnerships: The complex case of Bridgepoint Active Healthcare in Ontario, Canada. *Public Works Management and Policy* 1–22. https://journals.sagepub.com/doi/abs/10.1177/1087724X19899406

IMF-World Bank. (2023). *Debt Sustainability Framework for Low-Income Countries*. February 2023. www.imf.org/en/About/Factsheets/Sheets/2023/imf-world-bank-debt-sustainability-framework-for-low-income-countries

Inderst, George. (2013). Private infrastructure finance and investment in Europe. *EIB Working Papers*, No. 2013/02. European Investment Bank. https://econpapers.repec.org/paper/zbweibwps/201302.htm

Inderst, George. (2019). Infrastructure at the crossroads. *GLIO Journal*, Issue 04. www.glio.org/_files/ugd/a4e9c5_63978ffb139b48218f4b5c367d18defc.pdf

Inderst, George. (2021). *Private Capital Mobilization and Institutional Investors*. https://papers.ssrn.com/sol3/papers.cfm?abstract_id=3806742

Indian Infrastructure. (2021). *Bond Market Update*. December 2021. https://indianinfrastructure.com/2021/12/28/bond-market-update/

International Finance Corporation (IFC). (2005). *UN Global Compact and Federal Department of Foreign Affairs. Investing for Long-Term Value: Integrating environmental, social and governance value drivers in asset management and financial research*. October 25, 2005. www.ifc.org/wps/wcm/connect/9d9bb80d-625d-49d5-baad-8e46a0445b12/WhoCaresWins_2005ConferenceReport.pdf?MOD=AJPERES&CACHEID=ROOTWORKSPACE-9d9bb80d-625d-49d5-baad-8e46a0445b12-jkD172p

International Monetary Fund (IMF). (2015). Press Release: *International Finance Institutions Announce $400 Billion to Achieve Sustainable Development Goals*. Press Release No. 15/329. www.imf.org/en/News/Articles/2015/09/14/01/49/pr15329

IPE Real Assets. (2022). Fund management industry resumes fast-paced growth. *IPE Real Assets* (magazine). July/August 2022, Pp. 56–62. https://realassets.ipe.com/top-100-infra-managers/top-100-infrastructure-investment-managers-2022/10060812.article

Jablonski, Adam and Marek Jablonski. (2020). Business models in water supply companies: Key implications of trust. *International Journal of Environmental Research and Public Health*.

Jacobs, Karrie. (2022). Smart cities: Toronto wants to kill the smart city forever. *MIT Technology Review*. June 29, 2022. www.technologyreview.com/2022/06/29/1054005/toronto-kill-the-smart-city/

Jaggi, Rohit. (2018). *Investment Risk Management: Defining the Right Framework*. Strategi. www.stradegi.com/app/uploads/2018/04/Investment-Risk-Management.pdf

Jansen, M. (1989). Water supply and sewage disposal at Mahenjo-Davo. *World Archeology* 21(2)177–192. www.jstor.org/stable/124907

Kell, George. (2018). The remarkable rise of ESG. *Forbes*. July 11, 2018. www.forbes.com/sites/georgkell/2018/07/11/the-remarkable-rise-of-esg/?sh=5f45edd51695

Khumalo, Mlungisi Jimmy, Ireen Choga, and Elias Munapo. (2017). Challenges associated with infrastructure delivery. *Public and Municipal Finance* 6(2). www.researchgate.net/publication/319564292_Challenges_associated_with_infrastructure_delivery

Kliesen, Kevin L. and Douglas C. Smith. (2009). Digging into the infrastructure debate. *Regional Economist*. Federal Reserve Bank of St. Louis. www.researchgate.net/publication/227438319_Digging_into_the_infrastructure_debate

Knoepfel, Ivo. (2004). *Who Cares Wins: Connecting Financial Markets to a Changing World*. The Global Compact. www.unepfi.org/fileadmin/events/2004/stocks/who_cares_wins_global_compact_2004.pdf

Kozlowski, Rob. (2021). Milliman: Largest public pension plans' funding ratio hits 85%. *Pension and Investments*. www.pionline.com/pension-funds/milliman-largest-public-pension-plans-funding-ratio-hits-85

KPMG. (2016). *Foresight: A Global Infrastructure Perspective*. https://home.kpmg/xx/en/home/insights/2021/04/foresight-global-infrastructure-perspective.html

Lam-Frendo Maria. (2021). *If you issue it they will come: Lessons from recent infrastructure Bonds*. Global Infrastructure Hub. www.gihub.org/articles/lessons-from-recent-infrastructure-bonds/

Lavanchy, René. (2020). Greenfield infrastructure: investor appetite goes from red to amber. *IPE Real Assets*. https://realassets.ipe.com/infrastructure/greenfield-infrastructure-investor-appetite-goes-from-red-to-amber/10045388.article

Leach, Anne, Carmen Aguilar Garcia, and Sandra Lavuille. (2022). Revealed: more than 70% of English water industry is in foreign ownership. *The Guardian*. November 30, 2022.

Lipshitz, Clive and Ingo Walter. (2019). *Bridging the Gaps: Public Pension Funds and Infrastructure Finance*. NYU/Stern. https://ssrn.com/abstract=3319497

London Summit. (2009). *Leaders' Statement. Group of Twenty*. www.imf.org/external/np/sec/pr/2009/pdf/g20_040209.pdf

LOPRESPUB. (2016). *Public Infrastructure in Canada*. Hill Notes. Library of Parliament. January 25, 2016.

Lucas, Deborah and Jorge Jimenez Montecinos. (2021). *A Fair Value Approach to Valuing Public Infrastructure Projects and the Risk Transfer in Public Private Partnerships*. National Bureau of Economic Research. www.nber.org/books-and-chapters/economic-analysis-and-infrastructure-investment/fair-value-approach-valuing-public-infrastructure-projects-and-risk-transfer-public-private

Macquarie Asset Management. 2021. *Putting the World's Long-Term Savings to Work*. September 2021. www.mirafunds.com/au/en/our-approach/our-funds.html

Manulife Investment Management. (2021). *Tech and the Consumer have Transformed the Future of Emerging Markets*. www.manulifeim.com/retail/ca/en/viewpoints/equity/Tech-and-the-consumer-have-reshaped-todays-emerging-markets

Marshal & McLennan. (2018). *Infrastructure Asset Recycling: Insights for governments and investors*. www.marshmclennan.com/insights/publications/2018/jul/infrastruct ure-asset-recycling-insights-for-governments-and-investors.html

McKinsey & Company. (2017). *Reinventing Construction: A Route to Higher Productivity. Executive Summary*. www.mckinsey.com/~/media/mckinsey/business%20functions/ operations/our%20insights/reinventing%20construction%20through%20a%20produ ctivity%20revolution/mgi-reinventing-construction-executive-summary.pdf

McKinsey & Company. (2017). *Voices on Infrastructure: Transforming Project Delivery*. www.mckinsey.com/~/media/mckinsey/business%20functions/operations/our%20i nsights/voices%20on%20infrastructure%20transforming%20project%20delivery/voi ces-on-infrastructure-transforming-project-delivery.pdf

McKinsey & Company. (2022). *Charting New Approaches to Capital Project Delivery*. www.mckinsey.com/capabilities/operations/our-insights/global-infrastructure-ini tiative/voices/voices-on-infrastructure-charting-new-approaches-to-capital-project- delivery

McLaren, Duncan and Julian Agyeman. (2015). *Sharing Cities: A Case for Truly Smart and Sustainable Cities*. MIT Press.

Melas, Dimitris. (2019). *The Future of Emerging Markets*. MSCI. www.msci.com/ documents/10199/239004/Research-Insight-The-Future-of-Emerging-Markets/

Mercer. (2020). *Infrastructure Investing: Private Market Insights*. www.mercer.com/cont ent/dam/mercer/attachments/global/gl-2021-infrastructure-a-primer.pdf

Mercer. (2021). *Global Pension Fund Index 2021: Pension Reform in Challenging Times*. Mercer CFA Institute Global Pension Index | Mercer Canada

Mercer. (2021). *Dry Powder Meets Low Interest Rates–Time for a Private Market Boom or Bust?* www.mercer.com/content/dam/mercer/attachments/global/investments/gl- 2021-dry-powder-in-private-markets.pdf

Mintzberg, Henry. (1987). Crafting strategy. *Harvard Business Review*. https://hbr.org/ 1987/07/crafting-strategy

Mintzberg, Henry. (1996). Managing government: Government management. *Harvard Business Review*. https://hbr.org/1996/05/managing-government-governing-man agement

Mitchell, Patrick and Matthew Job. (2019). The rise and fall of PPPs. *Infrastructure Investor, The Decade*

Multinational Development Banks and Development Finance Institutions. (2021). *Mobilization of Private Finance*. January 2021. www.ifc.org/wps/wcm/connect/ 8249bfb4-2ad0-498d-8673-90fe196cb411/2021-01-14-MDB-Joint-Report-2019. pdf?MOD=AJPERES&CVID=ns1zGNo

Nature. (2021). Financing adaptation. Nature Climate Change 11:887. https://doi.org/ 10.1038/s41558-021-01213-4

O'Brien Isobel. (2022). GIC, OTPP, AIMCo to fund the world's largest green energy platform. *Infrastructure Investor*. June 17, 2022.

OECD. (2015). *Infrastructure Financing Instruments and Incentives*. www.oecd.org/fina nce/private-pensions/Infrastructure-Financing-Instruments-and-Incentives.pdf

OECD. (2018). *Amounts Mobilized from the Private Sector for Development*. www.oecd. org/dac/financing-sustainable-development/development-finance-standards/mobil isation.htm

OECD. (2020). Resilient and sustainable finance. ESG and institutional investment in infrastructure. *Business and Finance Outlook 2020*. www.oecd-ilibrary.org/finance- and-investment/oecd-business-and-finance-outlook-2020_eb61fd29-en

OECD. (2021). *Green Infrastructure in the Decade for Delivery: Assessing Institutional Investment.* www.oecd.org/env/green-infrastructure-in-the-decade-for-delivery-f51f9 256-en.htm

OECD. (2023). *Private Finance Mobilization by Official Development Finance Interventions.* January 2023. www.oecd.org/dac/2023-private-finance-odfi.pdf

OECD/G20. (2014). *Base Erosion and Profit Shifting Project*, OECD. www.oecd-ilibr ary.org/taxation/oecd-g20-base-erosion-and-profit-shifting-project_23132612

PEW Charitable Trusts. (2021). *The State Pension Funding Gap: Plans Have Stabilized in Wake of Pandemic.* www.pewtrusts.org/en/research-and-analysis/issue-briefs/2021/ 09/the-state-pension-funding-gap-plans-have-stabilized-in-wake-of-pandemic

Pickrell, Don H. (2022). The development of evaluation methods for infrastructure projects. In: *Infrastructure Economics and Policy: International Perspectives.* Jose A. Gomez-Ibanez and Zhi Lui (Eds.). Lincoln Institute. Cambridge, Massachusetts. Chapter 6, pp. 143–173.

PPP Council. (2016). *2016 Canadian Infrastructure Report Card: Informing the Future.* Canada Council for Public-Private Partnerships. www.pppcouncil.ca/web/News_Me dia/2016/2016_Canadian_Infrastructure_Report_Card__Informing_the_Future.aspx

Prahalad, C. K., and Hart, S. L. (2002). The fortune at the bottom of the pyramid. *Strategy and Business* 26:54–67.

Preqin. (2016). The $1 bn club: The largest infrastructure fund managers and investors. *Real Assets Spotlight.* Preqin. https://docs.preqin.com/newsletters/ra/Preqin-Real-Assets-Spotlight-December-2016.pdf

Preqin. (2016). *Preqin Quarterly Update, Q1 2016.* https://docs.preqin.com/quarterly/ inf/Preqin-Quarterly-Infrastructure-Update-Q1-2016.pdf

Preqin. (2021). *Global Infrastructure Report 2021.* www.preqin.com/insights/global-reports/2021-preqin-global-infrastructure-report

Primack, Dan. (2022). *Blackrock Strikes Back at ESG Critics.* Axios Pro Rata. September 8, 2022. www.axios.com/2022/09/08/blackrock-strikes-back-at-esg-critics

Probitas Partners (2019). *Infrastructure Institutional Investor Trends: 2019 Survey Results.* https://3asstpm1ai412ap5q1o60dzh-wpengine.netdna-ssl.com/wp-content/ uploads/2019/09/probitas_partners_Infra_Survey_2019.pdf

Rogers, Lucy. (2018). Climate change: The massive CO_2 emitter you may not know about. *BBC News.* December 17, 2018. www.bbc.com/news/science-environment-46455844

Rosen, William. (2010). *The Most Powerful Idea in the World: A Story of Steam, Industry and Innovation.* Random House.

Rozenberg, Julie and Marianne Fay. (2019). *Beyond the Gap: How Countries Can Afford the Infrastructure They Need while Protecting the Planet.* Sustainable Infrastructure; Washington, DC: World Bank. © World Bank. https://openknowledge.worldbank. org/handle/10986/31291

Sartori, Andrew. (2006). The British Empire and its liberal mission. *Journal of Modern History* 78

Sawant, Rajeev. (2010). *Infrastructure Investing: Managing Risks and Rewards for Pensions, Insurance Companies & Endowments.* Wiley.

Schrag, Zachary. (2002). *"Urban Mass Transit in The United States".* EH.Net *Encyclopedia, edited by Robert Whaples.* http://eh.net/encyclopedia/urban-mass-tran sit-in-the-united-states /

Schroders. (2019). *ESG Policy Risk & Impact Assessment Report: Infrastructure Finance.* December 2019. https://prod.schroders.com/fr/sysglobalassets/global-ass ets/french/aida/191219-schroder-aida-esg-risk--impact-assessment-report---decem ber-2019.pdf

Shackleton, Robert. (2013). *Total Factor Productivity Growth in Historical Perspective.* U.S. Congressional Budget Office. Working Paper 2013-01. www.cbo.gov/sites/defa ult/files/113th-congress-2013-2014/workingpaper/44002_TFP_Growth_03-18-201 3_1.pdf

Smith, Gregory. (2019). The promise of infrastructure. *Infrastructure Investor: The Decade.* September 2019.

Stanley, Martin. (2019). Towards a sustainable future. *Infrastructure Investor. The Decade.*

Task Force on Climate-Related Financial Disclosure. (2022). www.fsb-tcfd.org/about/

Thinking Ahead Institute. (2021). *Global Pension Funds Assets Study – 2021.* www.thi nkingaheadinstitute.org/research-papers/global-pension-assets-study-2021/

Tian Yifeng, Zheng Lu, Lu, Tian, Peter Adriaens, R. Edward Minchin, Alastair Caithness, and Junghoon Woo. (2020). Finance infrastructure through blockchain-based tokenization. *Engineering Management* 7(4):485–499.

Tomer, Adie, Joseph W. Kane, and Caroline George. (2021). *Rebuild with Purpose; An Affirmative Vision for 21st Century American Infrastructure.* Metropolitan Policy program at Brookings. April 13, 2021. www.brookings.edu/essay/american-infrast ructure-vision/

Tulchinsky, Theodore H. (2018). *John Snow, Cholera, the Broad Steet Pump; Waterborne Diseases Then and Now.* National Library of Medicine, NIH. www.ncbi.nlm.nih.gov/ pmc/articles/PMC7150208/

U.S. Department of Transportation. (2020). *Value Capture.* Centre for Innovative Finance Support. www.fhwa.dot.gov/ipd/value_capture/

UN-Habitat. (2008). State of the World's Cities 2009/2009. https://unhabitat.org/state-of-the-worlds-cities-20082009-harmonious-cities-2

United Nations Climate Change (UNFCCC). (2022). *Introduction to Climate Finance.* https://unfccc.int/topics/introduction-to-climate-finance

Verma, Shashi. (2022). New technologies in infrastructure. In: *Infrastructure Economics and Policy: International Perspectives.* Jose A. Gomez-Ibanez and Zhi Liu (Eds.). Lincoln Institute of Land Use Policy. Cambridge, Massachusetts.

Vogl, Bernard and Mohamed Abdel-Wahab. (2015). Measuring the construction industry's productivity performance: Critique of international productivity comparisons at industry level. *Journal of Construction Engineering and Management* 141(4). https://researchportal.hw.ac.uk/en/publications/measuring-the-construction-industrys-productivity-performance-cri

Wahba, Sadek. (2021). Commentary: Pension funds can launch a new infrastructure era. *Pension and Investments.* www.pionline.com/industry-voices/commentary-pens ion-funds-can-launch-new-infrastructure-era

Watts, Jonathon and Elle Hunt. (2018). Halfway to the boiling point: the city at 50°C. *The Guardian.* August 13, 2018.

Weber, Barbara and Hans Wilhelm Alfen. (2010). *Infrastructure as an Asset Class; Investment Strategies, Project Finance and PPP.* John Wiley & Sons. United Kingdom. P26.

Weisman, Jonathon and Bradley Graham. (2006). Dubai Firm to Sell U.S. Port Operations Move to End Three-Week Dispute Comes After GOP Lawmakers, Defying Bush, Vowed to Kill Deal. *The Washington Post.* March 10, 2006. www. washingtonpost.com/archive/politics/2006/03/10/dubai-firm-to-sell-us-port-operati ons-span-classbankheadmove-to-end-three-week-dispute-comes-after-gop-lawmak ers-defying-bush-vowed-to-kill-deal-span/bf9a9e42-3dce-469f-934c-922ac747c85e/

Woetzel, Jonathan, Nicklas Garemo, Jan Mischke, Martin Hjerpe, and Robert Paltre. (2017). *Bridging Global Infrastructure Gaps*. McKinsey Global Institute. www.mckinsey.com/capabilities/operations/our-insights/bridging-global-infrastructure-gaps

World Bank. (2020). *Urban Development Overview*. www.worldbank.org/en/topic/urban development/overview

World Economic Forum. (2020). *Global Competitiveness Report Special Edition 2020: How Countries are Performing on the Road to Recovery*. www.weforum.org/reports/the-global-competitiveness-report-2020

Yescombe, E.R. and Farquharson E. (2018). *Public-Private Partnerships: Principles of Policy and Finance*. Elsevier: Academic Press.

Zhang, Xueqing. (2009). Win-win concession period determination methodology. *Journal of Construction and Engineering Management* 135(6):550–558. www.researchgate.net/publication/245284092_Win-Win_Concession_Period_Determination_Methodology

Index

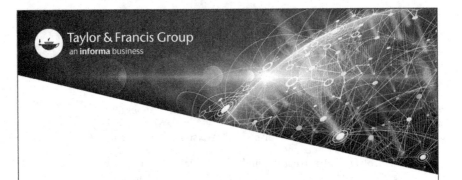

Printed in the United States
by Baker & Taylor Publisher Services